The Captain's Log: From Conway and Clan Line
to Trinity House

Captain James Alexander Riach

The Captain's Log
From Conway and Clan Line to Trinity House

James Alexander Riach

with an Introduction by Glen Murray
and an Afterword by Alan Riach

The Grimsay Press

Published by:

The Grimsay Press
An imprint of Zeticula Ltd
The Roan
Kilkerran
KA19 8LS
Scotland
http://www.thegrimsaypress.co.uk

First published in 2013

ISBN 978-1-84530-139-2

He resembled a pilot, which to a seaman is trustworthiness personified.

Joseph Conrad, *Heart of Darkness*, Chapter 1

Acknowledgements

Thanks are due to Aileen Soğuksu, Cem Soğuksu, MacArthur Cunningham, Carl MacDougall and Glen Murray, and to all those who have cared for James Riach in his later years.

The front and back cover paintings are both of the Royal Terrace Pier on the River Thames at Gravesend, by John Cunningham, reproduced by permission.

Contents

Illustrations

Introduction

Some years ago my old friend Alan Riach came sailing with me on the West Coast of Scotland.

We visited several places about which Alan was particularly curious; not only for their inherent historical or literary interest, but because of associations they had for his father.

Reading Alan's father James Riach's account of his childhood those connections and associations take shape and life and fit into a fascinating story.

Alan asked me to write a few words of introduction to Jim's book because of my connection with the sea and I'm very honoured to do so.

I shared Jimmy's fascination with the sea and, like him, followed the impulse to earn a living (though on my part only in part) on it. I shared with him the fascination of the element itself and the adventurousness of crossing it to visit foreign shores.

There, to some extent, our common ground ends. I became a yacht delivery skipper and spent many years sailing yachts of all types and ages to wherever their owners wanted them, a job that took me to some very out of the way places and into the company of some pretty unusual people. The literary interests that I shared with Alan got me involved with the literature of the sea which resulted, among other things, in my editing an anthology entitled *Scottish Sea Stories*.

And there, Jimmy and I reconnect. For his is a fascinating sea story. An account of a maritime career and way of life about which, until I read his book, I confess I knew almost nothing, an area of maritime experience that I haven't found elsewhere in the literature of the sea.

Jimmy's career in the Merchant Navy in the years after World War 2 makes absorbing and entertaining reading but it is his account of his career as a Trinity House Pilot that was new and particularly interesting.

Although the only time I've been skipper of a vessel that required a pilot was when transiting the Panama Canal (where even the smallest yacht has to carry a Canal Pilot) I have watched pilot cutters in all sorts of conditions in ports in many parts of the world carry these enormously skilled and dedicated men to and from ships of all shapes and sizes. The sight was so familiar that I never thought to any great extent about what their daily work actually involved.

Jimmy Riach's stories brought the work of the Pilot to life in a way I haven't come across anywhere else. His stories, often understated but full of detail, made me laugh and gasp as well as revealing the complex and exacting work that was the daily lot of the Thames River pilot in decades gone by and, I dare say, of Pilots the world over.

Though as vital as it has always been to the safety of shipping, in recent years the job of the Pilot has changed with changes in shipping technology and practice. Ships are now larger, there are fewer of them and they don't go to a lot of the places they used to need a pilot to get to. Jimmy's book provides a fascinating insight into the Pilot's role at one of the high spots in its history but, as well as being a valuable documentary, it's also a warm and absorbing personal story.

Glen Murray

1. Family and Childhood

My father was born in Dingwall, in Easter Ross, but as far as I know, my grandfather and great grandfather came from the Elgin area and had some sort of connection with either gardening or landscaping. I've a suspicion that one of my forebears was a gardener to an estate in that area because the house in Dingwall where we stayed, the house of my parents, was called Urrard, which was the name of part of an estate in the Elgin area.

James Riach at Urrard

My father's father was the Station Master at Dingwall and he retired from the Station Master's house, which of course he had to vacate, and moved into Urrard, the house my father had bought. He must have been in his sixties or seventies then. This is the man I was named after. A fine man, an upstanding man and a pillar of society, so to speak. The Station Master of Dingwall. He always seemed to me to be a formidable figure, from a small boy's point of view, and no-one questioned his authority in any way. He appeared to me to be somebody to look up to and to take advice from. I can't remember any specific times or specific incidents, it's just a general impression I have of him. He was very kind to me always.

James Riach's grandfather

I can remember my grandparents quite distinctly. My grandfather on my father's side was James Alexander Riach and on my mother's side was John Hood and I can remember both of them very well. I'm a bit more vague about my grandmothers but I still remember them.

My father, Patrick Scott Riach, was in the Post Office Service. My mother was Christine Laurie Hood and she came from Wick. My mother's father was a monumental sculptor with a pretty flourishing business in the north. The Hood family business had branches in Thurso, Stornoway and Dingwall. Their business is sold now. I think it's still under the name of Hood but no one in the family is connected with it now.

Hood monumental sculptors

My parents, I suppose, were really quite good examples of their generation. They were pretty straightforward people. My father was a pretty fine character, as was my mother. They didn't seem to me to have any bad qualities at all.

James Riach and parents

My father was a volunteer during, or just after, the beginning of 1914-18 war, and he was sent to Fort George near Inverness for initial training as a private in the army in one of the signals divisions. He had a good deal of communications experience, through his Post Office service. He was later transferred to the newly formed Royal Flying Corps, went to France and was promoted on the field to Second Lieutenant. He was promoted on the field again to Lieutenant and finally to Captain. When he came back from France he went for training, I believe in Norfolk, in King's Lynn or somewhere near King's Lynn, then eventually he went with the expeditionary force to North Russia under General Ironside.

Group of four officers and sergeants in the snow in Russia, Patrick Scott Riach, James Riach's father, second from left.

Air Force Cross

He had a very adventurous time up there and had difficulty in getting back. I don't think he came back until some time after the war had ended. He was awarded the Air Force Cross for his work in the Archangel area with British expeditionary force.

Meanwhile my mother was at home. She also had worked in the Post Office and was in the Dingwall Post Office at the same time as my father and that's where they met, after the First World War.

I was born in 1928 in Dingwall and the first years of my life were spent moving around Scotland. For the most part, my sister Jean and I were brought up in the Western Isles of Scotland, in Tobermory and Stornoway. My father left Dingwall when I was two to take a position in Dunbar. He was second to the Head Post Master and that's where I started my schooling. From there, probably when I was about eight or nine, we moved to Tobermory on the Isle of Mull, which at that time was a Head Post Office, controlling the islands of Eigg, Rum, and Muck, part of Morvern, and all of Mull, the whole area was under the Post Office at Tobermory. So that was another promotion for him as well.

Tobermory, Isle of Mull

Tobermory was home. I never travelled to the other islands at that time, but occasionally, when my father checked the sub-offices around Mull, he'd take me with him. Sometimes the whole family went across to Morvern in a hired boat and he would check the Office over there and we would have a picnic and things like that. But he never took us to the other islands. He would visit the other offices, to show the flag, so to speak. I don't think he was really checking up to see if they were honest or anything like that. He was just making sure things were running okay and if they had any problems and so on. It was an annual thing.

James Riach and sister Jean

So my father, his father and the Hoods formed a pretty solid professional middle class background. My maternal grandfather, John Hood, in particular was a pillar of the establishment. He was certainly one of the earliest and eventually one of the oldest members of the Boys' Brigade, which he was interested in all his life, having been a compatriot and colleague of William Alexander Smith, who started the Boys' Brigade and came from Thurso.

He was a staunch member of the Church of Scotland, but I don't think he was ever on the town council, though he might have been for he was greatly respected in the town. Hood Street in Wick is named after him. His photograph is in Wick Heritage, the local museum, and a considerable display is dedicated to his work, both for the town and the Boys' Brigade. I was made Lance-Corporal in the Boy's Brigade in 1943, when I was fifteen or sixteen. I still have the certificate somewhere, I think.

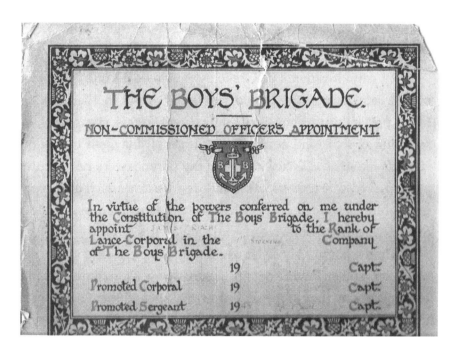

Boys' Brigade certificate

Despite the fact that Riach is a Gaelic word – thought to have derived from someone with grey hair – there was never any trace of Gaelic referred to or mentioned in a family context. Never. We never spoke Gaelic. My father and mother and my grandparents as far as I know had no Gaelic at all, but beyond that I couldn't say. There must have been Gaelic speakers at some point, maybe further generations back. There was never anything spoken about Gaelic, about the language or about the name Riach. I can't recall any direct reference to language in either of the families, the Riachs or the Hoods.

And both were fairly literary; in fact, my father's father founded a literary society in the railway when he was Station Master, and a very good society it was. It was a discussion group, a reading group. They met frequently, presented papers and discussed literary matters and characters, Burns, naturally, and they were interested in the literary figures of the day. I think I have some papers connected with the society. I haven't referred to them for some time so I can't remember exactly what they contain.

But I cannot remember Gaelic ever being referred to. And I cannot recall any of the family, my father or my mother, talking to me particularly about literature, writers, artists, composers, or anything of that kind at all. It was unusual for them to talk on subjects like that. Discussions with my parents were not lengthy or deep. We were a very close family and we loved each other very much, but we were never people who sat down and discussed lots of subjects among ourselves.

We had a piano. My sister could play the piano, not very well, and I remember having violin lessons for a short time. I can't remember why they stopped, but I certainly had them for a short time. Didn't do me any good.

(2) Should Canvassing for Parliamentary Elections be prohibited.

(3) Is an age of intellectual culture unfavourable to the development of great men.

(4) Are Trades Unions in their operations beneficial to Working men.

(5) Ought a member of Parliament to vote according to his own opinion or to that of the majority of his constituents.

(6) Parliamentary Debate.

(7) Can the Irish be justified in rising to action against the laws of the kingdom as they now do.

(8) Is Novel reading more productive of good or evil.

(9) Does ignorance or intemperance contribute most to crime.

(10) From which Season of the year do we derive the greater pleasure Summer

Literary Society Report: Topics for Debate

My own early reading was, I think, *The Skipper*, *The Rover* and *The Wizard*. I can remember reading all these 'penny dreadfuls' which I enjoyed very much but of course, I graduated to *Treasure Island* and things like that and the school put me in a direction of reading books of perhaps a more profitable nature, especially once we got to Tobermory.

I just don't remember when I began to read properly apart from school, apart from the books that I had to read for school. I was good at English and I was good at Art and Geography and History. I wasn't very good at Mathematics but I struggled along. I wasn't all that hot on languages either, but I did very well at the general subjects. And there was seeing foreign places on the map.

As I've said, we went from Dingwall to Dunbar when I was about two, in about 1930, and then from Dunbar to Tobermory a bit later, perhaps the mid-1930s, and then from Tobermory to Stornoway, then from Stornoway to Wisbech in Cambridgeshire. At least my parents went to Wisbech, but by that time I was away. So I was in Tobermory when I was between about eight and thirteen, then in Stornoway between about fourteen and fifteen. I left Stornoway when I was fifteen and a half or so, to go to the training ship, HMS *Conway* for two years, before going to sea.

Always, from an early age, I was aware that I wanted to travel, I wanted to see the world. And then of course, having been brought up in the Highlands and Islands, talking about foreign places was something that people did, because so many of the men and the sons of all the families were at sea as deep sea merchant sailors. It was a livelihood they could combine with their crofts and it was quite common for people from the islands to go to sea. It was a natural process.

I was curious about the world. I wanted to get away and explore the different places I'd read about, to find out for myself what they were like. And of course I was intensely interested in ships. Africa, India – these places fascinated me, and when we went to the cinema, we saw cowboy films and places like Arizona, New Mexico and Texas became fascinating.

It was a great event when the cinema came to Tobermory. Everyone looked forward to it. The films were shown in the Town Hall and we all took cushions to sit on. It expanded my horizons. I wanted to see everything, but Africa was the place that fascinated me most. And the sea fascinated me. And ships fascinated me. From as far back as I can remember I wanted to go to sea in ships, even before we went to Tobermory I was interested in ships, though none of my forebears had been at sea, except, on my mother's side, I understand that somewhere in her relations there were sailing ship captains. But the war had a big bearing on it of course. But yes, I like ships. I've always liked ships.

Before the war, cruise liners came into Tobermory Bay. And there were always other ships, like the Fleetwood trawlers who called in after fishing in Iceland. And they were always a source of interest to boys. We'd go out to them sometimes in an old rowing boat and get prawns and chatter with the fishermen who always welcomed us. We knew some of the skippers. But when the war came the whole scene changed and it all got very exciting and adventurous. So I'm sure that was how my fascination started; though, of course, we had men leaving Tobermory to go to sea and some of them didn't come back.

I remember the Campbell family in Tobermory. The father was a captain with Burns and Laird and his son and son-in-law were both seafarers. Well, his son, Reggie Campbell, was second mate in Hogarth and Co's Baron Line and we knew the family very well. He was lost early in the war, lost at sea, and that was a big shock to me, a big shock because I knew him.

Strangely enough, as far as I know, only two of my contemporaries at school or any of my school friends went to sea.

They were the sons of the *Lochinvar* captain, a MacBrayne's boat that sailed between Tobermory and Oban. I was in touch with them for some time after they'd gone to sea, but I've lost touch with them now. There weren't many of my contemporaries who went to sea. They all went to higher things!

I remember, my first sight of the sea was at Dunbar. That's the first memory I have of seeing the sea. I had probably seen it before that but I can't remember.

And when I was about ten, I remember falling into the sea off Tobermory pier. I was fishing for cuddies. The boys always fished and fishing for cuddies was a general pastime. We had a line and a hook and a bit of rope on the end of it, and fished off the pier for these small fish, which were inedible but fun to catch. I was at the end of the old stone pier with a friend of mine, just the two of us. The tide was fairly low, but we were on the bottom of the stone steps leading down to the end of the pier, fishing happily, and I got too enthusiastic, went over too far and went straight into the water. I can remember vividly seeing the bubbles going up and my eyes were wide open as I went down and I couldn't swim but my friend, he stretched out something, I can't remember what it was, a stick or something, and he pulled me in. I clambered onto the pier and squelched my way home, where I was severely reprimanded by my mother, of course.

My sister was there too of course. Jean was three years older than me. She had a better brain than me and was better looking, so she was well known and people remembered her fondly. She became a nursing sister in the Glasgow Royal Infirmary. But our age difference was just enough to keep us that bit away from each other. Her interests were not my interests.

I don't remember ever having a holiday. The war interfered with all that and the islands were a restricted area. You couldn't move about anyway, although we regularly went back to Dingwall to see my grandparents and we spent a week or two there and then a week or two in Wick to see my mother's parents. So all the holidays I remember were in Dingwall and Wick. I think I must be more of a Highlander than a West Coast man, although Tobermory is really, in a way, home.

I think Tobermory probably had the biggest influence on me. There is an attractiveness about Tobermory in the shape of the place and the character it has, and in a way it's still there, and it's there in some of the characters that I knew. But on the other hand, my experience of it was also terribly heightened by the fact that it was in the context of the Second World War.

War broke out when we were in Tobermory; and when it became a very important naval base, Tobermory changed out of all recognition. Suddenly, there were hundreds, thousands of sailors, and this was a great shock to the community. And it opened up horizons to boys of my age.

Everything around was so interesting. The horrors of war just didn't enter into the picture, except through adventure stories in *The Hotspur, The Skipper* and so on. We were aware of all of this, but it didn't touch us in a horrific sense. It was all a great adventure for boys with all these ships around, all these sailors. It had a big impact on me.

The Tobermory base was commanded by Sir Gilbert Stephenson. He had pioneered anti-submarine techniques in the First World War and Tobermory was the Royal Navy's Anti-Submarine Training School from 1940 till the end of the war.

At one time there were maybe fifteen or twenty ships in the bay, all training for anti-submarine work. And we even had an air raid, a very tame affair compared with London, Glasgow or Coventry. I don't think they could find Tobermory. They went round and round and dropped their bombs in the hills and killed a few sheep.

Then there was the time when the commandos were training in Ardnamurchan, across the Sound of Mull, and they decided to have an exercise raid on Tobermory. Stephenson considered it was no concern of his and the result was that the commandos came in on the landing craft and opened fire. Then the ships opened fire. The people of Tobermory didn't know what was happening. They thought that a submarine might have come into the bay.

I was asleep and my father woke me up and I could hear the explosions. We thought we were being bombed. He said, 'You better get up. We better take cover here.' So we disappeared underneath the stairs with the dog and my mother. And I thought my last day had come.

It turned out the commandos were Canadian. And when the local police turned out to combat them, the commandos arrested them, took over the Mishnish Hotel as their headquarters, locked the police in their rooms and ran riot through the town. It was really upsetting. They did a great deal of damage to the shop fronts. Many had their windows broken and there was glass everywhere.

The commandos had a good time, but for the Tobermory community that one night showed them war was a serious business.

But there were others who carried on as usual, men like Dykes. He was a great Tobermory character. I suppose you could call him a likeable rogue, but a rogue he certainly was. He thought nothing of authority and bucked it whenever he had the chance. He was really an anti-social type, but at the same time one always had a sneaking liking for him. He served time in prison, for being a vagrant I think, but he felt that the policeman who arrested him had been particularly vindictive and when he came back to Tobermory and was walking down the gangplank, the policeman was waiting for him and said something like, 'That'll teach you!'

and as soon as he did that, Dykes took a swing at him and knocked him down and he was sent back to prison. That was the story I remember. He was a wild man. When he was drunk, the police had to look after him. I think he froze to death in the open, and was discovered by a ditch at the roadside, with his hair frozen into the grass. He was only one, but there were many of them there in those days, characters.

From Tobermory we went to Stornoway on the Isle of Lewis in the Outer Hebrides. Experiencing different parts of Scotland allowed me to meet a variety of people. The Scots of these places differed, and also there were the Canadians and the Fleetwood trawlermen, the Dutch and French, who came on the submarines mostly sent, I think, to rest for a couple of weeks and to take part in additional training.

But the people I met, of whatever nationality and language, were ethnically European. I never really encountered people from other races. I don't think I ever saw a black face. But we had Italians in Tobermory. The Capaldi family. They had a shop selling ice cream and confections and so on, and there was a Pakistani family in Stornoway. They had been there for a very long time, long enough to be Gaelic speakers, which was good for trading, good for business.

Apart from that, I had no connection with anyone of a different coloured skin until I went to sea. And all that's part of the story because it's part of the nature of what travel was about in that era. The nature of meeting people of different races was tied up with the nature of going to foreign and exotic locations. The lure, the mystery of Africa was also to do with the fact that the people of such countries were foreign to my experience. But of course, nobody travelled much in those days unless you were very rich. Travelling abroad would almost always be related to your work. There was no tourist industry, not as we know it today. No-one could afford to go abroad. And then of course the war restricted everything. And these were the years when I was growing up.

On the other hand, there was always a sense of Scotland as a coherent identity. I was always aware of being Scottish. I always thought that the Scots were the finest nation in the world. And I was always extremely proud to be Scottish, always. From an early age that was absolutely true and a natural thing. It wasn't anything anyone said or anything I was taught. I just assumed that I was Scottish and I was the greatest. Sounds very egotistical, but I was young.

When I went to Stornoway I was in the same school and must have been in the same class as the writer Iain Crichton Smith. But I never talked to him because of the prejudice between Gaelic speakers and English speakers and boys from the town and boys from the villages. There were local distinctions and if he was indeed in the same class as me – and we were of an age, at the same school, the Nicolson Institute – it's reasonable to assume that's why we didn't meet. The country boys left in a bus as soon as the school finished, so we had no contact with them outside school, and even in school they kept very much to themselves and spoke Gaelic amongst themselves. And the town people, the town boys of Stornoway, not many of them really did have Gaelic. A number of them did, but quite a lot of them didn't, so language was a dividing line. There were town boys and there were country boys.

I can't remember Iain being in the class but then there were many others I can't remember being in the class either. We certainly both recollected the same teachers who belted us, when we met, many years later. We had the same teachers and were there in the same era. There's no question about that.

It was quite a hard time in the Nicolson Institute. It was a wonderful school at that time, one of the finest schools in Scotland. The teaching staff were first class. Looking back, I think they were very fine teachers and very clever men. But they were strict disciplinarians who hammered knowledge into boys without a doubt. And girls for that matter.

But I was always looking for ways to get away to sea. And of course the war was on and I was coming up to an age when I felt that I should be away and doing my bit.

Adventure was calling and I'd heard through the grapevine about HMS *Conway*, a training ship for merchant naval officers who accepted boys at fifteen and a half, and I thought this would be a good move. So, after a great deal of discussion, I eventually persuaded my parents that this was what I wanted to do. And a boy from Stornoway was already on the *Conway*, so we called on his experience. It was a considerable shock, an effort for my parents, not only financially, because it was a fee-paying thing, but emotionally, to allow me to go to the *Conway*. It was based in North Wales, which is a long way away from Stornoway. And it was a bit of an upheaval in the family when I did go, but go I did!

By this time, my sister Jean was working as a nurse in the Glasgow Royal Infirmary, so when I left home, my parents were alone. I left when I was fifteen and a half, or sixteen perhaps. I'm not sure, somewhere around there. I had never warmed to Stornoway and Lewis as I did to Mull and Tobermory. I was older, of course, when I went to Stornoway, and though it had its influence on me, it never really took me to its heart as Mull did. But I've been back to Stornoway on several occasions, been back to Lewis and enjoyed it, but not as frequently as to Tobermory. But it was no great wrench when we left Mull to go to Stornoway. It was just a great adventure.

The most influential part of Stornoway was the fishing fleet. There was an enormous amount of fishing going on out of Stornoway at that time, herring fishing. And I was interested in this and this was a great social pleasure to me.

But to leave home was a great adventure. I was going on my own. And it was like Scott setting out for the Antarctic! That would have been 1944 perhaps. The war was still on. I had already been to Glasgow and Edinburgh. They were just enormous great cities, which I didn't particularly like and was always glad to get away from them. We had relations in Glasgow and we visited them before the war. And then when we were living in Dunbar we weren't far from Edinburgh and there were relations in Edinburgh too. So we did visit these places but it was only to visit relatives. Sight-seeing was not on the itinerary. There was simply no such thing as sight-seeing as such. I had no interest in the cities.

But now I was leaving Scotland altogether, and for the first time. I was about to head through England to North Wales. This was a crucial part of my life. It was a turning point. It was an opening up of everything for me. Imagine it: a young boy setting off, having to travel for three days then donning the uniform of an officer cadet in the Royal Naval Reserve. To get through the travelling and arrive in this completely new environment was just a total shock, absolutely stunning.

I found myself in a world I had no idea existed. I was with people I'd never encountered before, boys of so many different interests and different backgrounds. We had boys from the Empire, the West Indies, Bermuda and so on, all British. You had to be of British parentage, otherwise you were not eligible, so we were all British.

I'd never really encountered English people before I went to the *Conway*, except among the sailors who came into Tobermory but I'd never met anyone English who was my own age. There were the Fleetwood trawlermen, but this was just a totally foreign environment. To this day I don't know how I managed to cope with all this, but you're very adaptable when you're fifteen and a half or sixteen. And there were other Scots on the ship. You had no friends when you arrived there but you made friends, very deep friendships. And the whole procedure of training was a very short period of time, two years, but I think probably that two years had the biggest influence on my life, more than anything that's happened, before or since.

2. Apprenticeship and Training

So it was two years spent on the *Conway*, 1944 and 1945, except for a month or so at the Outward Bound School in Aberdovey.

The Aberdovey deck boys training school was started by Alfred Holt, who owned the Blue Funnel line in Liverpool. The intention was to give lads a training in sea survival skills. We learned small boat work, lifeboat work, swimming, physical exercise and initiative. It was a month of hard graft, but the result was that if we had to depend on our own resources, we at least had some training behind us; and the month at Aberdovey was the only time I left the *Conway*.

The *Conway* was a wooden warship launched in 1839 as HMS *Nile*. She had survived the Baltic Blockade in the Crimean War and had protected British possessions in the Caribbean. In 1876 she was renamed *Conway* and moored on the Mersey. She was moved to the Menai Straits when Liverpool was bombed and for 250 of us this was home for two years, cadets in training to become officers in the Royal and Merchant Navies.

It was essentially a fee-paying public school. All the usual subjects were on the curriculum, with an emphasis on the maritime subjects of navigation, seamanship, signaling, ship construction and stability. You were under a type of naval discipline and for twenty-four hours a day were trained to be an officer. And a very fine training it was. I believe it stood me in good stead all my life.

We wore Royal Naval Reserve Cadets uniform, blue reefer jackets, brass buttons and naval officers' caps. The Cadet Captains had gold braid. It was all very naval and traditional and I quite enjoyed that. Of course, I missed home, but it was so absorbing that you didn't really have a lot of time to be homesick. You had to look after yourself and every minute of the day was regulated and busy.

HMS *Conway*

I was being instructed on how to give orders rather than take them, even though the training mostly involved taking orders. We were gradually given more responsibility as we proceeded through the regime. The more senior you became the more responsibility you got and by the time you left you had a lot of authority.

There were six terms, and you proceeded from the first term through to the sixth. When you went onto the second term there was a new intake in the first term, when you went onto the third term there was another first term intake, so that during your stay cadets were continually coming and going. When you reached your sixth term, you were the top of the heap and could give orders to everybody.

There were very few officers, perhaps six or seven and they had petty officers, but essentially there were only six or seven officers in charge of 250 boys, so they passed all responsibility for the discipline onto the boys themselves. And in your sixth term you were either Senior Hand, a Cadet Captain or a Senior Cadet Captain. There were Senior Cadet Captains and Junior Cadet Captains and they had a lot of authority.

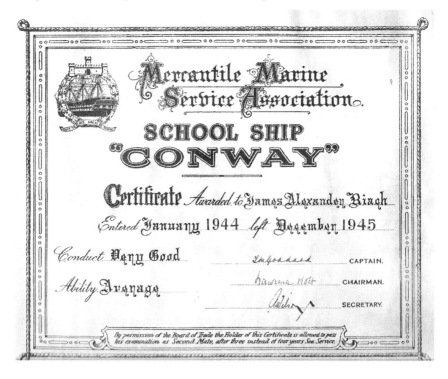

HMS *Conway* Certificate

The whole operation worked very well indeed, but the discipline was extremely harsh, very strict, as it had to be, because we were all pretty lively lads. Corporal punishment was quite common. We worked with tarred rope as they did it in Nelson's day. The rope ends were called teasers and they were used to beat you on the backside, just to give you a little flick here and there to keep you going. They stung, but never drew blood. Never, even though they were painful.

But for serious offences such as stealing or smoking or blatantly breaking the rules, it was a formal beating, not in public, but we all knew when it was going to happen. The whole ship went dead quiet and it was done in a very serious way. It would never be tolerated today. But we knew where we were, if we broke the rules we took our punishment and that was it.

But we did break the rules, of course. I did. Most of us did at one time or another. We were always hungry, and purloining, I won't call it stealing; simply liberating food from the stores was perfectly normal as far as the boys were concerned. And I was discovered, which resulted in me being punished, but it didn't bother me too much.

The food was absolutely diabolical, as a result of the war, of course, and the food shortages. We could have had naval help, but for some reason this was turned down. As I understand it, the *Conway* was offered naval ship status, which would have meant naval rations and they would have been far better than the civilian rations we were subject to, even though the cooks were Blue Funnel Line Chinese cooks. They were all right but the food was the problem, mostly dried egg and ration grub, though it's also true that we were young healthy boys who were always looking for more. In retrospect it may not have been as bad as I remember, but most of the cadets depended a lot on food parcels from home.

There was a man on the *Conway* who thought the world was flat; an old man who had spent his life at sea, probably couldn't read or write and was working as a general seaman on the training ship. He was an old sailing ship bosun who was working on deck above a classroom where the cadets were being taught geography. The skylights were open, he heard the teacher and obviously disagreed with what was being said. Now, I wasn't in that class. This didn't happen to me, but the story was told to me by a cadet who swore it was true. In the middle of the lecture the old bosun shouted down through the skylight, 'The world isn't round! The world is flat! It's not round, it's flat! I know it's flat! It can't possibly be round!' The poor man was causing great hilarity.

Conway logo

I think the term lasted three months, but it was a school, so we had school holidays, not quite the same, but we did have fairly long holidays. I went home at regular intervals. It involved a three-day journey. We normally broke up on a Wednesday, but I was the only cadet who was allowed to leave on the Sunday previously and was the envy of the whole ship. But it also worked against me, because I had to leave home three days before everyone else to get back on time.

I travelled from north Wales to Crewe on the Sunday and remember having great feeds in the Forces canteen and going to the cinema, which was packed with service people and smoke everywhere. We watched the old classic Hollywood films and there always seemed to be a good film on.

I went to pass the time before catching the night train to Glasgow, but it was always a pleasurable experience. The trains were crowded, people sleeping in corridors, all full of service personnel and so on. And being in uniform, of course, I was able to enjoy all the facilities of a serviceman, the canteen facilities, the WVS coming round with cups of tea, that sort of thing.

I arrived in Glasgow too late to catch the morning train to Inverness. There were only two trains a day and the night train was next, so I spent the day with my sister, Jean, and arrived in Inverness on Tuesday morning. Again, I had to spend most of the day in Inverness because the train for Kyle of Lochalsh left in the afternoon. I'd catch the boat on Tuesday night and arrive in Stornoway about three o'clock on Wednesday morning! But it didn't bother me. I enjoyed it. It was fun. It was an adventure.

I was a boy when I went to the *Conway* and think I was probably still a boy when I left, but I was very much more in command of myself, able to face anything that was coming, and ready to kick off. It made a big difference to me. I must have been seventeen and a half or eighteen when I left and by that time the war was over. Some of the cadets went straight to the Royal Navy Reserve during the war and were consigned to the Royal Navy later on. They were the minority; the majority of boys went directly into the Merchant Navy. And they could pick their company.

I chose the Clan Line, simply because it was Scottish and was known as the Scottish Navy. It was a big, well-known, well-established, respectable shipping company whose ships serviced a span of routes and that appealed to me. So I applied to the Clan Line and they took me immediately.

I have to say I had no regrets about leaving the *Conway*. I couldn't get away from the place quick enough. It was only in later years that I realised the value of those years.

I remember the whole thing very clearly, especially particular individuals. I can remember all the officers. They had an enormous influence on me. The individual officers were great characters. They were all old navy men. The Captain was a naval commander and I think every boy in the ship respected him. The Chief Officer was always a Merchant Navy Chief Officer and the two Chief Officers who were there during my time were both very fine men.

And there were the other boys. We had a lot of friends and we formed great friendships, some of which I've kept up till this day. Indeed the connection with individual people was very important. Our officers had something and must have been well-suited for the work because they certainly had a lot of influence on me and on all the cadets. I know going to reunion dinners and talking with cadets, theirs was a long-term influence. They were all fine men.

It's history now, but I think discipline of that kind is of great benefit to boys who might otherwise go off the rails sometimes. It gives them an incentive, reason and hope and encourages them to do something with their lives, but I can't imagine people like that are given an opportunity to influence boys nowadays.

HMS *Conway*, Christmas 1945

3. The Clan Line

By the time I left the *Conway*, my parents had moved down to Wisbech in Cambridgeshire for my father to take up his last position before he retired. So that was where I went to wait for my call to the Clan Line, which came pretty quick, after a couple of months, I think.

I joined my first ship, the *Clan MacAulay*, in Birkenhead. I think that was in April, 1946, and again my whole life was completely turned over.

The *Clan MacAulay* in Tasmania

So here I am – at last on board a ship. This was wonderful, a great adventure – I'm going to sea – I'm a cadet and this is MY ship and I'm going away into these distant foreign places!

The *Conway* of course was never at sea. She was always moored, so I was a sailor who had never actually been to sea, except in small sailing ships back in Tobermory and Stornoway. I did go out with the fishermen, overnight or for a day or two at a time, for two or three days, sometimes. But that was fishing. We never called in anywhere except maybe to Mallaig. I think we nearly went to the Faroes once and that was a great adventure. But that was just out to fish and back into Stornoway. We were out of sight of land, of course, but this was different, this was the real thing.

The *Clan MacAulay* was a refrigerated ship. She carried frozen meat and I thought she was magnificent. Mostly, we ran general cargo to South Africa and East Africa, and then we crossed the Indian Ocean light-ship to Australia and changed crew on the homeward run.

We visited Cape Town, Port Elizabeth, East London, Durban, Lourenco Marques – now known as Maputo – and Baira, both in Mozambique. Then, we crossed to Australia, loaded meat in the Australian ports, went to India to change the crew in Cochin – now Kochi – then home through the Red Sea, the Suez Canal, the Mediterranean and back to London. I think it took about five and a half months.

She was entirely manned by an Indian crew, Lascar seamen. Only the officers were European. This was my introduction to a different culture. And it was just extraordinary, especially coming from the *Conway* discipline to go on board a ship where I sat at the saloon table, was waited on, and considered an officer. This was an extraordinary experience. And the food was magnificent. After the *Conway* it was like being in the Savoy or something. The whole experience was enormously pleasurable!

We had curry and rice you couldn't dream about. For breakfast we had probably one, two, three, four, oh, five courses! There was always a choice. Fruit juice, cereal and porridge, fish and kippers, and bacon and eggs and sausages and then toast, bread, marmalade, rice cakes, practically anything you wanted! If you didn't like the curry you didn't have to eat it, but it was always there and it was considered very lavish for that time. We were still under war time restrictions.

We sailed from Birkenhead to Cape Town. As soon as we left, a gale caused the cargo to shift, so we had to call into Falmouth to sort it out. Then a member of the crew was sick and we had to call in to Lisbon. This was my first foreign port, but all I saw of it was the lights coming into the Tegus. We were in and out in no time at all. So Cape Town was the first foreign port I saw in daylight.

James Riach, late 1940s

All this, of course, was totally new and very exciting. I remember looking at Cape Town and thinking this is just like a film! At that time coloured films were in their infancy, and it looked just like one of these coloured films. Everything was so bright and the atmosphere was so clear; the colour of the sea, the land, the buildings and the brilliance of the sunlight, was wonderful, wonderful.

It was a time of year when the weather was semi-tropical and as we progressed up the coast, the further north we got, nearer the equator, the hotter it became and much more tropical, though when we crossed the Indian Ocean into the southern latitudes the weather was very bad and cold, but by the time we arrived at Australia it was different again.

Cape Town was where I had my first introduction to a dead man. On the morning we arrived, one of the stewards, a Goanese, was discovered dead in his bunk. I knew nothing about this until the second mate said, 'I need you, now! Come on, you're coming with me.'

I followed him out and there was an ambulance on the quayside at the bottom of the gangway. 'Come in here,' the second mate said. 'I need a hand to carry this man ashore. You take his legs, I'll take his head.'

They didn't even tell me he was dead until the second mate said, 'Just lift him. He's dead. It'll be all right.'

I was holding his feet thinking, 'He's dead. This man is dead. This man is dead'

I'd never seen a dead man before. It made an impression I haven't forgotten. I think the third mate gave us a hand and we carried him out covered in a blanket.

He had died of natural causes. I think he'd had a heart attack during the night. It shook me a bit at the time but it was all part of this life. Whatever I was told to do, I did it.

I served all my cadetship on the *Clan MacAulay* and my job was very varied. We kept watches on the bridge with an officer of the watch, we checked the lifeboats and worked with the carpenter. On an Indian-crewed ship, we never worked directly with the crew, we were always on our own. But we did a lot of work that the crew did. It was all very exciting and I must say I really enjoyed it.

The *Clan MacAulay* in Sydney

There were only about sixteen Europeans in the ship and the full compliment was about 110. Even in the lowly position of cadet, as I was, I still had some authority. But I didn't give orders as such. I was not in a position to order people around, but there were times when it happened.

I did a three-year stint as cadet, teaching me how to be a Merchant Navy officer. Then I came ashore, did a short time at school and passed an examination for second mate. These examinations were anything but easy. We were tested on seamanship, navigation, ship stability, compass deviation and meteorology, electricity, nautical signs and everything pertaining to ships and seamanship. And there was a very strict oral examination, which lasted anything from an hour to two hours.

James Riach, Second Mate

I took my second mate's ticket in London because the family were still living in Cambridgeshire. But by the time I took my mate and master tickets my father had retired and moved back to Scotland, so I took these exams in Glasgow.

My parents retired in Dingwall and that was still home, though I wasn't there very often, because leave was not generous. After a six-month voyage if you got two weeks at home you were quite lucky. When you went on leave you never knew when you would be recalled. You were always sent on leave until recall. You could expect a telegram at any time telling you to rejoin the ship.

When I got my second mate's ticket, I was appointed third officer straight away and joined the *Hesperia,* part of the Houston Line, which was a Clan Line subsidiary. We were just trading generally, mostly to India.

At first, I wasn't all that keen. The only experience I'd had of India was changing the crew in Cochin and I wasn't impressed. It was dirty and smelly and I didn't like it, but that was superficial; I was only in there for a night or a day, at most.

I was third mate of the *Hesperia* for eighteen months and that introduced me properly to India. We traded there a lot and I began to realise there was a little more to India than the smell and teeming masses. I began to enjoy it, especially in the north-east monsoon period when navigation was a joy because of the continuously beautiful weather. I became quite attached to India, although I remember feeling that I would have liked to have gone back and seen other, perhaps more civilised ports.

I left the *Hesperia* to go for my first mate's ticket in Glasgow and stayed in Bath Street at the Officers' Club, which is now gone. It was an institution, a place of great character and usually inhabited by characters, all up for their tickets of some kind or another. In particular, I remember Tommy McLeod who was a native of Wick and an employee of the Hogarths – H. Hogarth and Sons of

Glasgow – 'Hungry Hogarths' the infamous Baron Line, a tramp ship company. Tommy McLeod had come up from the decks. He had a struggle with his exams, but he got through them all. He and I were up for the same tickets and we lived in the same place and we struck it off and we had many adventures together in Glasgow. That was a time of high jinks!

When I got my first mate's ticket, I became second mate with the *Clan Macbeth* and to see through my time till I could sit for my master's ticket I was second mate on the *Clan Macbeth* and the *Clan Buchanan*. I went back to Glasgow for my master's ticket and was appointed straight away chief officer on the *Clan Kenneth*.

Master Mariner's Ticket

From then on there was only one rank left and that was master. I knew it would be a good number of years before I could progress, so I went from the *Clan Kenneth* to the *Perthshire*. I was chief officer of the *Perthshire* and then chief officer of the *Clan Malcolm*.

The *Perthshire*

The *Clan Malcolm*

I decided I couldn't wait to be promoted to master. It would take too long. It was a perfectly straightforward matter of waiting for dead men's shoes. I could see there would be no prospects for quite some time. And by then I was married.

Jimmy and Janette engaged

I met Janette when I was up for my master's ticket, through mutual friends, and it developed into a romance. We were married after my first voyage as chief officer and set up home in Hardgate, above Clydebank, on the shoulders of the Kilpatrick Hills.

That first voyage was very eventful. We went to Australia with general cargo and had arranged the marriage for when we got home, but the voyage took longer than we expected. We had a suicide on board and all sorts of problems in Australia, and the wedding had to be postponed.

It was my chief steward, the purser, who took his own life. He took laudanum from the medicine locker.

It was seven o'clock in the morning. The cadets came up to the bridge to get their work for the day and at the same time the purser's steward came up saying he couldn't wake the purser and he thought something was wrong.

I sent a cadet down to see what was going on and he came up in a panic. He told me I should go down, he didn't like what was down there at all.

I called the captain on the voice pipe. He was shaving. I asked if he'd come up and take the bridge so I could find out what was going on. He came up straight away. I went down and discovered the man was dead.

We thought there might have been a spark of life in him and organised respiration. We did what we could for about an hour, but by that time there was no doubt. It was a confused situation and quite traumatic.

We didn't have a doctor on board, so we radioed for any ship with a doctor in the vicinity to contact us. An Indian destroyer answered and she was close enough to rendezvous with us. They sent a doctor over, he certified the purser was dead and left us.

We buried him immediately because of the hot weather. The carpenter and I sewed him up in the traditional fashion with fire bars at his feet. The bars came from the engine room and were used for the boilers; they were weights, to make sure he sank.

It really wasn't very pleasant; we had a bottle of whisky on the table while we were doing it and that helped a bit. And we had a traditional burial at sea.

This caused problems when we got home because the matter had to be investigated, but, thankfully, the investigation didn't interfere with our wedding.

We had various changes of orders on that trip. We had very bad weather, the cargo shifted and the journey to Australia was quite dangerous. The weather was so bad we had to hove to for about three days and were unable to move in any direction. We were nearly caught in a typhoon, but thankfully it blew itself out before it caused us any danger. Then we had all sorts of trouble with the dock labour in Australia. They were always going on strike for no reason at all.

But back we came and it all was a bit of a dream for me because we arrived in Tilbury docks, I think on Easter Sunday, and we had difficulty getting a shipping master to clear me off the ship but finally it was done and I was married on the Tuesday, so I didn't have much time to change my mind. Didn't even have a clean shirt. But it all went very smoothly.

We were married in Glasgow University chapel and that was quite something. Janette was a school teacher and the girls from her class appeared at the wedding and gave us a cheer, very joyful of them. We had our honeymoon at the Marine Hotel, which is now the Regent Hotel, in Oban. I had two or three weeks leave and then I was off, back to sea.

I joined the *Perthshire*, a refrigerated ship and new to me. I was on her for two years or so. She was built to carry meat but she was getting on in years by the time I joined her.

Jimmy and Janette married

She was no longer suitable to carry frozen cargos, but we took chilled cargos, mostly fruit from South Africa, though we loaded in East Africa as well. She was a fine ship and I really enjoyed my time on her. I got on well there. I have very happy memories of the *Perthshire*. We did about six trips to South Africa for fruit, which was quite an easy job for me. We got the ship in good shape and things ran very smoothly, apart from the normal run of problems like typhoid and typhoid carriers.

James Riach, Chief Officer, the *Perthshire*

We had a fire as well, which didn't please us very much. It was in the hold. We loaded in East Africa and loaded a combustible cargo, even though the ship wasn't really suited. We didn't have the ventilation. It was crazy to load this kind of cargo and we told them, but I suppose the cost of the freight was high.

We did our best to keep the place ventilated but eventually spontaneous combustion set the stuff alight in the middle of the hold. It could have been extremely dangerous but we had anticipated the problem, battened everything down and hoped it wouldn't cause any dangerous problems until we got into port.

We were bound for Avonmouth and when we got in, the fire-fighting people came down and we discharged most of the cargo, but then we found that most of what was in the centre of the lower hold, Number 4 lower hold, was still burning slowly and ready to burst into flames. Had that happened, it would have blown up the ship, and really caused a lot of problems. But it worked out all right; just a normal occurrence.

We were away for about six months at a time. The timetables and the routes of the liner companies generally were pretty much set. We knew where we were going, we weren't tramping around looking for cargo. We had set runs, which was much better because we knew when we would be in the various ports and when we would be home. Tramp ships didn't know where they were going to be or where they were going and sometimes they were out for two years at a time. I never did a voyage of more than seven months.

After the *Perthshire* I was appointed to the *Clan Malcolm* for her second voyage. She was a new ship, a beautiful ship, but the captain was a big bully.

He was a fine seaman but he couldn't deal with people. He had no idea of how to handle men or handle a crew, especially a young crew, which we were. He was dodging on retirement and thought it was best to run the ship with iron discipline. I certainly didn't care for that because I got all the flack. I was the middleman, standing between everybody else and him; I did the hard graft and all he did was criticise and moan and bully, which caused the junior officers great distress and made my job very difficult.

I don't think it was a case of personal vindictiveness. The root of the problem was that he was one of the old style masters who thought things should be done in the old ways. Times were changing and I was a young mate with my own ideas, more modern ideas and I put them into practice. He didn't approve of some of them and I didn't approve of some of the things he wanted done. Anyway, he used to go into fits of rage for nothing, at stupid unpredictable things. Yet he had a fine war record and was a fine seaman. I have no criticism of his ship-handling or his seamanship. He was a very experienced man who had simply outgrown his time. He was retired early. It made him furious. He didn't want to retire. But they couldn't get anyone to sail with him. Agents around the world, stevedores, the people he had to deal with complained about him and the Clan Line decided he should retire, they said, to make way for younger men, but it was clear they'd had enough of the trouble he was causing. I lasted longer than any mate and was with him for something like eighteen months. We parted with mutual consent. But we nearly had a mutiny and, of course, we were together, shoulder to shoulder, when that happened.

The Indian crews – who never gave any trouble – were changing. India was beginning to assert its independence and insisted that instead of crews coming from what is now Pakistan, as they used to, we should take Hindus who were trained in India. The Indian Government forced the Clan Line to take these people, many of whom were Communists who didn't conform with the way ships are run and it was a cause of severe trouble.

Besides, to mix Hindus and Muslims in the same crew was asking for trouble. And the old bosuns, the Lascar serangs, and the petty officers who were in charge of them, were Muslim, while half the crew were Hindus. The old guys just didn't get on with them and there was a lot of trouble.

They said they weren't being given enough stores and attacked the second steward with a knife and that was a serious problem. The captain immediately called for help from a warship, but we had the situation under control by that time. I really disagreed with him sending for a warship, it was over the top altogether. He cancelled that and told them that we were okay.

There was an enquiry when we got home and I must say we were really shoulder to shoulder on this one and weren't going to let them get away with it. But they did get away with it, because the situation was political. The Clan Line didn't want to disturb the Indian Government because of their considerable trade with India. So the enquiry was a farce. We weren't censured at all. In fact the company congratulated us on our handling of the situation, but we wanted more, we wanted these people brought to justice for what they'd done. But they never were. The crew was changed and we never saw them again.

There was another conflict when we were discharging a general cargo in South Africa. Part of the cargo was whisky. We used to carry a lot of whisky, all over the world, and when we stopped work in the evening, we put tents over the hatches to keep the rain off. We didn't close the hatches because a lot of work was involved in closing off hatches, so we covered them with canvas tents. Fridge ships always carried these tents.

The third officer was going round, making sure that the tents were tied up properly, when he saw a hand coming up from the hold clutching a bottle of whisky. And he tried to grab the hand, but it slipped from his grasp. The bottle went one way, the hand went the other, the man disappeared down the hold and hid among the cargo.

When the third officer told me the story, I told him to get the cadets and go down to find him. Then, sitting in my cabin, I thought, 'This is not a good idea. This man might attack these guys. He could have a knife, or anything.' So I went down on deck and called them up.

We had Zulu watchmen on board, so I got the foreman and told him, 'Get your men. There's a man down in that hold who's been stealing cargo. I want you to go down and get him.'

'We will find this man for you,' they said.

Twenty minutes later there was a knock on my door, and the Zulu foreman said, 'We have caught the man.'

'Where is he?'

'We have manacled him to the rail.'

So I went out on deck to find this poor man manacled, beaten up and in a terrible state.

I asked the foreman, 'What have you done?'

'He is a criminal sir,' he said.

I sent for the South African stevedores' foreman who was a white South African. He was very upset.

'What's the next step?' I asked. 'I don't think we should go much further with this. What will happen if I don't call the police?'

'He'll never get a job on the docks again. He comes from a village and he's living in a compound. He'll be taken out of his compound, sent home and that'll be him. He'll never get a job down here again.'

In those days a lot of labour came from the inland African villages. They were put in compounds, stayed for nearly six months, earned some money and went home again.

Well, I thought to myself, he's been beaten up, that's enough punishment for anybody. I didn't intend to prosecute him. But to my dismay I found that if I didn't prosecute him the Zulu watch company would do so, probably for prestige reasons. I had no choice but to call the police and prosecute him, which caused us problems because we had to appear in court. I think the poor man got about six months hard labour.

Clan Malcolm, by James Riach

A long time later when I was chief officer on the *Clan Malcolm* and we were bound for Australia, we had loaded unmarked drums on deck. We didn't know what was in them and the manifest of the cargo which we always carried, which told us exactly what cargo we had on board, said nothing about the contents of these drums. Even after enquiring I was told we had to accept these drums as they were.

The Government told us to put them over the side, to discharge them as soon as we got over the 100 fathom line, which we did. We stopped the ship, rigged the derricks, and swung them over the side.

To this day I have no idea what was in them, but I have my suspicions. They were probably toxic material, some kind of dangerous material which the Government wanted dumped in the deep water. We were told to be very careful with them.

Then there is the story about the doughnuts and the monkey. It didn't actually happen to me. It was told to me by a pilot who had been a captain with a tramp ship company, an elderly man, very much older than me. I was a cadet and he was dodging on retirement.

He said that he'd been chief officer of a ship coming home from South America. They were painting the ship to get her nice and clean for arrival home. And the crew had bought a monkey in South America. Buying pets wasn't uncommon in those days.

Jacko used to trail around with the crew when they were painting, working on deck and so forth. The chief officer was on watch one afternoon. He was pacing up and down the bridge when he saw the monkey on the after end of the boat deck, sitting on the galley skylights. He seemed to be making actions with his hands, bending down into the skylight, lifting himself upright and then flinging his hand outwards. He repeated these movements, again and again.

The chief officer was intrigued. He watched for a while then suddenly the monkey gave an awful yell and disappeared up to the top of the mainmast. So when the chief officer was relieved for his evening meal, he thought, 'I'll go along to the galley and see what's been going on.' The cook was busy, but the chief officer asked what was wrong with the monkey.

'There was a lot of fun going on here half an hour ago,' he said. 'That damned monkey! I was making doughnuts in hot fat and hanging them under the skylight to cool off. I'd make a batch of doughnuts and suddenly they would disappear! Gone! I thought, there's something going on here. So I put up the next line of doughnuts and kept an eye on them. This hairy hand came down from the skylight. The monkey was picking up the doughnuts. They were quite soft but very hot, so he'd pull them free of the line, take a bite, but they were far too hot for him, so he'd spit it out and throw the doughnut over the side.'

The chief officer had seen the arm going down and the monkey throwing the doughnuts away. But the cook was so incensed about this that he flung the hot fat up to the skylight and the poor old monkey was scalded in about six places and disappeared up the mainmast for about two days and was almost bald when they finally got him down!

Poor old Jacko. They were a hard case crew, and a bit later on they stupidly left this monkey on a long lead on the foc's'le head when they were painting, while they went for a smoke. The monkey got in among the paint and scattered red lead all over the white paintwork and a week's work was spoiled.

The bosun was so incensed when he came back that he picked up the monkey and flung him over the side and he was never seen again!

The last end of Jacko. I think the sharks must have got him. They were a hard lot these crews.

I was told that story by a Captain Baron who was a pilot at Wisbech on the River Nene. He'd spent his life at sea, had been a captain for years before becoming a Trinity House pilot.

My father had his last post at Wisbech and I was home on leave, before going to sea and Captain Baron took an interest in me. In fact, I remember he said, 'What you want to do is to get your ticket and join the Trinity House pilot service, either in London or Southampton. That's the best job you could get.'

I'd be about eighteen then. But I used to accompany him, go with him down to the mouth of the Nene, to pick up the ships and bring them up to Wisbech.

Anyway, as I said, when I was on the *Clan Malcolm* I didn't get on with that particular captain and though I was perfectly prepared to stay at sea, I could see things were changing. The writing was on the wall. I can't claim to have foreknowledge or to even predict the demise of the British Merchant Service but something I didn't like was going on throughout the whole service.

And, of course, I was married now. I had a family and a son and I thought, if I carry on at sea, I would have to spend my life away from my family with very short spells at home and that didn't seem such a good idea, especially if I was going to have to wait for some years before I got a command of my own. I wanted it. If they had given me a command, I would probably have stayed.

But in the end I decided to apply to the Pilotage Service and I became a pilot on the London River.

4. Becoming a Trinity House Pilot

I didn't want to come ashore to an entirely new job, but I did want to have a normal home life, a family life, and I couldn't see this happening if I stayed at sea. So pilotage seemed a logical step.

I would have preferred to have been a Clyde pilot, so that we could stay in Scotland, and I applied for the Clyde Pilotage service, but I didn't know though I soon found that appointments were not entirely made on merit. It was a question of who you knew rather than what you knew, and my application form went into the Clyde and I never heard any more about it.

But at the same time I applied for the London Pilotage, and that was a different kettle of fish altogether. I was called for an interview before we sailed on the next voyage.

As well as being the General Lighthouse Authority for England and Wales, the Corporation of Trinity House of Deptford Strond, known as Trinity House, is also the official deep-sea pilotage authority, providing expert navigators for ships trading in Northern European waters. A court of thirty-one Elder Brethren, presided over by a Master, ruled Trinity House. The Elder Brethren were appointed from three hundred Younger Brethren, who were themselves appointed from lay people with maritime experience. I was interviewed by two representatives of ship owners, two pilots and two Elder Brethren.

And it just so happened they were recruiting pilots. There was a boom in London, I was placed on the list and waited my turn.

With perhaps eight or nine men before me, I was going to have to wait for maybe a year or more before I was called, so I didn't want to go away deep sea for too long.

I spent the last year of my time in merchant service relieving other chief officers on ships around the British coast and the continent. The Clan Line co-operated quite well, after a bit of arguing. In fact a superintendent told me they liked having some of their men in the pilot service, especially at London.

When I was eventually called, I had to spend six months training – at my own expense. The profession of a Trinity House pilot was so élite that there was no shortage of applicants. In fact, some people went to enormous lengths and considerable expense and weren't even successful. It was a prime, plum job, and you were expected to be trained at your own expense.

Trinity House didn't employ the pilots – who were self-employed – so they weren't going to pay you. And the ship owner certainly wasn't going to pay you. So you were left to your own devices for six months.

There wasn't much in the way of training; you just watched what went on and more or less trained yourself, with the help of the pilots who were going to be your colleagues.

Janette was teaching at a school in Airdrie, so we had an income there, but we had to cut down on everything. My son was with his grandparents, Janette's parents, at Calderbank, near Airdrie, while I went down to Gravesend, in Kent, and set myself up in digs where I was very well looked after. I had no problem there.

In fact, I had a great time, the time of my life. I could pick any ship I wanted to board, the pilots were very helpful and I enjoyed it thoroughly. I had a great time tripping and at the end I presented myself at Trinity House for an examination.

It was pretty thorough. You didn't have to be a genius but you had to have a very good memory and you had to really hammer at it to get all the stuff in because you had to know everything, everything about the river. And the Elder Brethren of Trinity House examined you very closely.

For example, there were a thousand buoys, maybe more, and you had to know all the characteristics, the depth of water under them, the courses and distances between them.

The big advantage pilots had was that although the examiners had the charts spread out in front of them, which we didn't have, they were not pilots themselves. And we not only knew the river, we knew that they didn't know it as well as we did.

The examiners continually referred to their charts when they asked questions. The Elder Brethren were quite used to this and had a set piece to go through. They examined me on everything that was on the chart. As far as practical pilotage, such as the handling of the ships was concerned, they assumed I had learned that during my tripping period, since there was no way they could have examined anyone on that. The only way you can be a pilot is to do the job. And it's difficult to examine you on the actual manoeuvring of a ship.

Anyway I was granted a license to pilot ships up to 14-feet draft, which was called under-book. I was an under-book man and that meant I was on half-pay because I couldn't pilot ships beyond 14-feet draft. It would be three years before I could earn as much as a fully-licensed pilot, a first class pilot, which didn't please me very much, but that was how it was done.

We weren't apprentices, we were pilots, though we were restricted in the ships we could pilot. But what caused me the most trouble has usually been the small ships rather than big ships. A big ship usually has better equipment, the officers are usually better, the whole bridge procedure is better and more efficient and you have a back-up system on a big ship. You've got power in the engines and usually, as far as the river pilots are concerned, they're in some ways more straightforward than some of the smaller ships, because some smaller ships went to some funny places in the river Thames. And we used to do things with smaller ships we wouldn't dream of doing in a big ship. Obviously, though, there were also times with big ships when your experience and your knowledge was tested to the limit.

I was an under-book man for three years, then I went for another license. We were licensed every year and examined every year, but only on the changes that occurred in the previous twelve months. The big examinations were when you were first licensed as an under-book man and three years later when you became what they call an All Draft First Class Pilot. And that was a big examination. But by that time, and I think most pilots felt the same, you were pretty confident. You knew you were going to pass because by that time you knew far more than the Elder Brethren did.

I was licensed as an upper-book pilot and off I went, whoof! Away! I could handle the biggest ships in the place and I was top of the heap, in the big money then! I really enjoyed that.

We brought our first brand new car and had our first foreign holiday, down to the south of France and we had a great time. By that time we had our daughter, and she came with us to France, while our son went back to Scotland to his grandparents. But unfortunately just as soon as we did that the dockers in London went on strike and if there are no ships there's no money!

Soon, though, we returned to work, and the next thirty years or so I spent as a Trinity House Pilot, working out of Gravesend, in Kent.

No. 2439

To all to whom these Presents shall come, The Corporation of Trinity House of Deptford Strond, send Greeting, know ye that in pursuance and by virtue of the powers given them for that purpose in and by The Pilotage Act 1913, and of all other powers them enabling, the said Corporation, having first duly examined *JAMES ALEXANDER BLIGH* of *52, WHITEMILL LANE, GRAVESEND* — Mariner, (the Bearer hereof, whose description is endorsed on these Presents) and having, upon such examination, found the said *JAMES ALEXANDER BLIGH* To be a fit and competent Person, duly skilled to act as a Pilot for the purpose of conducting Ships, sailing, navigating, and passing within the limits hereinafter mentioned, **Do** hereby appoint and license the said *JAMES ALEXANDER BLIGH* to act as a Pilot, for the purpose of conducting Ships from _____

And this Licence (if the same shall not be revoked or suspended in the meantime, as in the said Act provided), is to continue in force up to and until the 31st day of **January** next ensuing the date of these Presents, but no longer, unless renewed from time to time by Indorsement hereon. Provided always that the said Pilot shall so long comply with all the Bye-laws and Regulations made or to be made by the said Corporation.

This Licence shall not authorise or empower the said _____ *to take charge as a Pilot of any Ship or Vessel drawing more than* _____ *FEET WATER (except when an upper draught Pilot is not available to offer his services), in the Rivers* **Thames** *or* **Medway** *or any of the Channels leading thereto or therefrom, until it shall be certified hereon that the said* _____ *has acted as a Licensed Pilot for three Years, and has been on re-examination approved of in that behalf by the said Corporation.*

Given under the Common Seal of the Corporation of Trinity House of Deptford Strond, this _____ day of _____ 19__.

Deputy Master _____

Secretary _____

Pilot's Licence

5. The Trinity House Pilots of the London River

I imagine most people will have heard of river pilots, of pilotage and Trinity House. And though a dictionary definition of pilotage is 'a pilot's fee,' it can more specifically be defined as 'The safe handling of vessels in restricted and dangerous areas, including berthing and unberthing.'

Pilots were known to exist 4,000 years ago in the days of Abraham and are recorded as being at work in Ur of Chaldea. In those days Ur was a flourishing port, although today the city is far from the sea.

Pilots are mentioned in the Book of Ezekiel. And the Code of Hammurabi, named after the Babylonian king and one of the earliest deciphered writings of significant length in the world, written around 1700 BC, laid down the payment for pilots as two shekels of silver.

Pilots were used where and when they were needed but, apart from loose guilds and societies, often of a semi-religious character, there was no actual organisation. In the 12th Century, the Island of Oleron, which was a seafaring island in the Bay of Biscay and an important maritime community, laid down the Laws or the Rolls of Oleron, the first formal statement of maritime law in north western Europe and based on similar judgements which had governed Mediterranean commerce since before the time of Christ. They were introduced to England in the 12th century, to Flanders in the 13th century and are the basis of maritime law to this day.

The Laws of Oleron, among other things, stated, *It is established for a custom of the sea that yf a shyp is lost by defaulte of the Lodeman, the maryners may, if they please, bring the Lodeman to the windlass or any other place and cut off his head withoute the maryners being bounde to answer before any judge, because the Lodeman has committed high treason against his undertakynge of the Pilotage. And this the judgement.*

The term 'lodeman' came from the lodestone which had magnetic properties and which mariners in ancient times used as a compass and was commonly used to describe a pilot. The word 'pilot' comes from the Dutch 'pijl' meaning anything vertically straight, and 'loot' or 'leod' meaning sounding lead.

When the Laws of Oleron were introduced to England and incorporated in the Black Book of the Admiralty, beheading was replaced by hanging from the yardarm.

It is thought that Stephen Langton, who was Archbishop of Canterbury from 1207 until his death in 1228 and the man credited with dividing the Bible into the chapters we know today, founded a semi-religious Guild of Mariners who were prestigious enough to have a large hall and almshouses at Deptford Strond and it was from this guild that pilotage in England began to become an organised and regulated profession.

In 1512, a petition was presented to King Henry VIII for a Royal Charter. The reasons are given in part of the petition: *Wherfortyme owte of mynde as long as due order good rule and guyding were suffird to be had in your said ryver and other places by auncient English Maisters and Lodesmen of the same, the said ryvers and places and the daungers of the same were then by them thrughly serchid so surely that fewe shippes or noon were perished in defaulte of Lodemanage. Now it is so moste Gracious Sovereign Lord, that dyvers and many young men namyng themself Maryners, beying oute of all good order and rule not having the perfyte knowledge ne experience in ship mens crafte, neither of sufficiency experience approved ne of age in the same to know the suraunce and saufconduyte of shippes by the comying of lodemanage, daily, unseytly medle therewith to great hurte and loss of moche of the said navye.*

Another part reads: *Also that Scots, Flemings and Frenchmen have been suffered to learn, as lodesmen the secrets of the Kings streams, and in time of war have come up as far as Gravesende, and fette owte English shippes to the great rebuke of this Realm.'*

Henry was occupied with war with France, but the Charter was granted finally on 20 May 1514 when it gave official recognition to Trinity House, awarding general powers for the safety and progress of shipping. The Guild was to have the title of 'The Guild of Fraternitie of the most glorious and Blessed Trinitie and St Clement in the Parish Church of Deptford Strond in the County of Kent.'

There would be one Master, four Wardens and eight assistants. The first Master of the new Corporation was Sir Thomas Spert, a Deptford man who was captain of Henry's flagship the *Mary Rose* and later master of the *Henri Grace a Dieu*, the largest vessel then in England and Clerk-Comptroller of Henry's navy.

It is thought that, in forming the new corporation, Henry was, in part, influenced by what was happening in Seville, where Emperor Charles V of Spain had founded a lecture on the art of navigation and appointed Sebastian Cabot, son of the explorer John Cabot to be Pilot Major for the examination of Atlantic pilots. Henry later made this same Sebastian Cabot Grand Pilot of England.

The new organisation had many and varied duties, some connected with the Royal Navy but as far as London pilots were concerned, there was now a properly constituted body under royal charter to look after their affairs and protect their interests. The charter was confirmed by Edward VI in 1547, by Queen Mary in 1553, and by Elizabeth I in 1558, when Trinity House was given the responsibility for lighthouses, buoys and seamarks. James I granted a new Charter in 1604 and this was where the Elder and Younger Brethren are first mentioned.

Cromwell dissolved and disbanded Trinity House in 1647 on the pretext of the house having loyalist tendencies, but Charles II reinstated Trinity House in 1660. He also issued a new charter on 8 July 1685 and this was the basis of the present day constitution. The new charter ran to 16,000 words, largely due to the influence of Samuel Pepys who was the first Master of Trinity House to be elected under the new charter. Other Masters have included the Duke of Wellington.

Records of the activities of Trinity House are not as easy to find mainly because of a series of disastrous fires in 1666 and 1714. In 1940, Trinity House was blitzed and destroyed and with it, nearly all its priceless books, pictures, models and, of course, its documents and records which were packed in readiness for transportation to a place of safety which was to happen the very next day.

It is known that six years after the granting of the first charter, there were 40 licensed pilots from London to Gravesend and that no ships had been lost. And the first recorded licensed pilot, Ambrose Marshall, was licensed on 24 February 1696 and could work downriver as far as Gravesend.

From 1514 until 1988, Trinity House functioned as the leading pilotage authority. For almost 500 years, they examined, disciplined and recruited pilots for the London river and, most importantly, always provided adequate numbers of pilots of first-class quality to suit the shipping industry.

Trinity House, however, does not, and never has, directly employed pilots, although it does lay down the rates which are paid by ships for pilotage services. Pilots are self-employed and, generally speaking, were very individualistic and guarded their comparative freedom fiercely. In my time as a pilot, the day-to-day affairs of the pilot station, the running of some pilot boats and the methods of working, as well as many other administrative duties, were entirely handled by pilots with little interference by Trinity House. This relationship worked reasonably well. After all, the Elder Brethren and the pilots came from the same background, had the same sea qualifications and were often old shipmates and friends. There was a certain mutual respect, which at times became a little strained, but stood the test of time.

In my day, the system of payment was organised like this. At the end of a voyage, the captain would sign a formal chit the pilot gave him and the pilot would take that to the office of Stubbs, in Harmer Street in Gravesend, just down from the Clock Tower. Stubbs was the honest broker who would claim the payment from the ship's agents in accordance with the signed chit. The amount of that payment could vary greatly, according to the size of the ship and the company, and so on. But the money held by Stubbs was paid out to the pilots regularly, in accordance with their rank. We got a salary by sharing the funds in this way. It was a kind of democratic structure, a communist ideal in a strange way, and it worked well. It eliminated the hostility that might have been exacerbated if individual pilots had been competing for bigger salaries and bigger ships and made sure we all got a fair deal. Stubbs is long gone, now, of course.

Modern Trinity House has kept many of the old traditions and practises though, of necessity, it has had to move with the times. The House is still run by the Deputy Master and Elder Brethren, and they still wear frock coats on ceremonial occasions, as well as attending to the serious business of running the corporation on modern guidelines. Their interest in London pilotage affairs came to an end in 1988 when Government policy decreed in its wisdom that pilots would come under the authority of each individual port and become paid employees of the local harbour authority, so that today pilotage is controlled by the harbour authorities of the ports involved.

The vast majority of pilots were not impressed with the new order and fought it bitterly to the end, when many senior pilots opted for early retirement.

Thus ended an era and although Trinity House will continue to operate as the lights authority and maintains its other capacities, it will never be quite the same again.

Another factor should be touched on before leaving the history of pilotage, and that is the Acts of Parliament that have been passed purely in connection with pilots. From 1561 through to 1766, there were Acts passed concerned with sea marks, beacons and so on, and they presumably had a connection with pilotage. However, in 1766, an Act of Parliament officially brought into existence the Liverpool Pilot Services and then, in 1800, the Humber pilots made successful application for an Act to be passed to prevent illegal piloting and to control compulsory pilotage in the Humber. In 1890, after great deliberation, a new Act incorporating many far-reaching changes through the country was passed and amended in 1913.

There are probably fewer than 2,000 pilots in the British Isles today and to take up so much valuable parliamentary time for such a small number of men may seem strange and out of proportion. The fact is that when, once again, in 1988, the full procedure for presenting and passing a new Act was embarked upon, the result was that pilots were made more 'accountable' to the parliament in Westminster than probably anyone else in the country.

I left the pilotage service in the same year.

6. The Making of a Trinity House Pilot

How does a man become a ship's pilot? What qualifications are required? What sort of training is necessary?

In my own case, as I noted, the first step to a pilotage career began with two years' training on HMS *Conway*, from 1944 to 1945. Then at the age of eighteen, I spent three years as a cadet on the Clan Line and a further twelve years with the same shipping company, rising through all the ranks and gaining all my certificates of competency up to Master Mariner. To be eligible to become a London Trinity House pilot, I had to be under the age of thirty-five, the holder of a Master Mariner's certificate and, in practice, to have been Master or chief officer of a large ship. My application for a position as a Channel Pilot (the total area downriver from Gravesend) was accepted when I was thirty-two years of age in 1960. Then, armed with references, I had to attend an interview board consisting of two Elder Brethren of Trinity House, two ship owners' representatives and two serving pilots. Having been successful, I was then placed on a waiting list of candidates and was required to hold myself in readiness to take up the appointment when a vacancy occurred. If I was thirty-five or over before a vacancy cropped up, then the offer of a position in the Pilot Service was cancelled.

I successfully negotiated this hurdle and reported to Gravesend, where I spent the next six months, at my own expense, doing 'trips' (with a qualified pilot) as a trainee. On completion of this period, I presented myself at Trinity House before an Elder Brother, who had to ensure, through rigorous examination, that I had all the in-depth knowledge of the pilot district in order to be granted a licence.

I may add that Trinity House did not take into account the possibility that a candidate may be on the other side of the world, or at sea, when he was called for interview, or to start work. If a candidate did not show up as requested, his name was deleted from the list.

For the next three years, I was confined to piloting ships of 14' draught or less and was awarded a handsome vellum licence with an impressive red seal, and became an 'Under Book' or 'Under Draught' Pilot. The next step was another searching examination and the granting of an 'All draught licence' and, from then on, I was entitled to pilot any ship regardless of draught or size and was a fully-fledged 'Upper Draught' or First Class Pilot. Apart from an annual examination to ensure continued familiarity with the pilotage district and to check on eyesight and general health, the formal process of making a pilot had been completed. I personally qualified, by taking further examinations, for licences to enable me to pilot ships in the North Sea, Channel and round the coasts of Britain and the continent, but these additional qualifications were not compulsory. As can be seen, the procedure of making a pilot was protracted and reflected the popularity among the seafaring community of aspiring to be a Trinity House pilot. The service had a wealth of prospective candidates coming forward and that made recruiting very selective indeed.

After the procedure of selection and training and licensing, a pilot then slotted into a roster system according to his name, and a never-ending belt of pilots were called upon when necessary, depending on the volume of traffic coming down the river. Thus, a start was made to a varied and unusual way of life. Ships are no adherents to normal hours and a pilot could be called to work at any time of the day or night. Weekends and holidays meant nothing. Tides were more important than clocks, and sleeping and eating habits were anything but regular.

Trinity House was the Authority. Trinity House examined, licensed and disciplined pilots, but Trinity House did not employ the pilots. The large pilot cutters on the sea stations were manned and owned by Trinity House, but the pilot boats at Margate and Gravesend were run by pilots.

Pilot Cutter, Gravesend

The day-to-day running of the pilot station, including leave arrangements, duty pilots and so on, was all managed by pilots. A democratically elected 'General Purposes' committee attended to these matters with great efficiency. Working rules were set up by pilots and were strictly adhered to, and a cutter committee was in place to run the pilot boats. There was also the Pilotage Act of Parliament and the Bylaws of the Port of London which served the interests of the pilots as well as the port.

Pilots were classed as self-employed, a mixed blessing, but one which gave to pilots a freedom of action which they cherished in spite of the restrictions imposed by the Pilotage Act, local Bylaws and such like. Pilots were not allowed to work except within laid down parameters, pilotage fees were regulated and pilots were even subject to restrictions on where they could live and what clothes they could wear. They were not allowed to work when and how hard they liked, and severe penalties were in place for pilots who were incompetent, unreliable or otherwise unsatisfactory. However, the pilot's freedom of action on the bridge of his ship was total and he answered to no one except the Master of his ship. In other words, his work was totally concerned with the conducting of his ship and the safe and speedy conclusion of his act of pilotage.

7. The Royal Terrace Pier

For more than one hundred years, the Royal Terrace Pier had been the home of the Trinity House pilots based at Gravesend. The pier was the focal point, the key co-ordinate point. It provided shelter, information, food and companionship and was vital to the business and social life of the pilotage community. The pier dominated a pilot's life. Most pilots disliked it and yet also had an almost hypnotic respect for it. The pier fascinated them. They had a love-hate relationship with it which is difficult to express and almost impossible to understand. Even today, many pilots talk about the old pier, which vanished in the 1970s, with great affection and nostalgia, but with no regret for its demise.

The Terrace Pier was built in 1844 and was renamed in 1863 when Princess Alexandra of Denmark arrived here, *en route* to St George's Chapel, Windsor and her marriage to Queen Victoria's eldest son and heir, Edward, Prince of Wales.

The structure of the buildings on the end of the pier, which housed the river pilots on one side and Channel pilots on the other, was mainly wooden, draughty and basic, and was a type of hybrid shore fo'c'sle of a sailing ship and the outer office of a Glasgow tramp ship owner's business premises – all combined into one Victorian pillared finger which pointed across the river towards the Essex shoreline.

The outer area consisted of windows overlooking the river, a desk like a minister's lectern upon which a slate rested with the same confidence as a Bible in a parish church. Various odd items adorned the walls – an ancient mercurial sea barometer which was never consulted, a brass plate, kept highly polished, recording the names of pilots lost in the two world wars, notice boards crammed with information, most of which pilots had been familiar with weeks beforehand, a little alcove to make tea and a telephone. Various ancient settees were prominent, and an old carpet of sorts, with a well-defined centre strip worn almost threadbare by scores of pilots' feet as they paced up and down between window and door.

The inner sanctum was the 'pièce de résistance.' As I recall, there were nine open-ended wooden bunks, placed against the inner wall – end to end – with little curtains which could be closed. The result was that, when fully occupied, the bunks were designed so that the feet of one pilot were within inches of the head of the man in the next bunk. At one time, there had been a 'bogey' (an old-fashioned boiler) as a centrepiece to the room and, latterly, an oil-fired abortion was fitted at one end, both of which emitted the most obnoxious of fumes which, together with various odours of strong tobacco, sweating feet and general wind-passing, resulted in a rich atmosphere, especially at 2am on a busy winter's tide when every available sleeping space was occupied.

The pier had character – mostly bad and very powerful. It defied being taken for granted and was continually exerting its influence over the men who used it day and night.

Friday was usually an extremely busy day. If the high water was timed for the evening, ships were very often programmed to finish discharge or loading on the Friday night and sail before the weekend. The pier was then throbbing, with perhaps 20 or 30 pilots on standby.

There was a constant flow of men in and out, the air would be thick with tobacco smoke, and arguments and discussions mainly about shipping and pilotage affairs were the order of the day.

The Royal Terrace Pier and boats on the Thames, by John Cunningham

Anchors were metaphorically dropped all over the place, and 'full asterns' and 'hard a starboards' were common terms used in the general hubbub.

When a large, individually-minded group of men are thrown together in such intimacy, there is always camaraderie, humour, a little friction and a great deal of talk, and the pier did have problems. The sleeping arrangements alone would have defied the best efforts of the most expert time-and-motion study man. When a bunk or settee was vacated, there was almost always another pilot ready and waiting to take up the place, and well do I remember the gratitude I felt for the warmth and comfort bequeathed to me by a colleague who had just been hauled out to serve some Dutch *schuyt* on her way to sea on a dirty winter's night.

Very often, when a pilot was No 1 and on turn, he had very little time to leap out of a bunk and board his ship (one notoriously heavy sleeper swore that he was never fully conscious until he shook hands with the master of the ship). It was not uncommon for pilots, in their haste, to dress in another man's gear, particularly shoes and caps. The result of these mistakes can be imagined when a pilot with small feet dives into a pair of shoes too large for him, and leaves his own smaller shoes to create agony for a man of larger dimensions. Most of these difficulties were amicably overcome and articles of clothing returned to their rightful owners in due course.

The old pilot station has long since gone, but the pier itself is still used by pilots to embark and disembark from the pilot cutters. The new pilot station, opened by the Duke of Edinburgh in the 1970s, is much more comfortable, fitted with electronic aids and with a comfortable lounge and sleeping accommodation of a high standard. However, in common with many modern buildings, the atmosphere is quite different from the old pier and has naturally affected the pilots' attitudes towards their headquarters. No pilot would tolerate a return to the conditions of the past, and yet we all agree that something good has been lost by the change and that whatever has gone, has gone forever.

8. *Ternefjell, 1962*

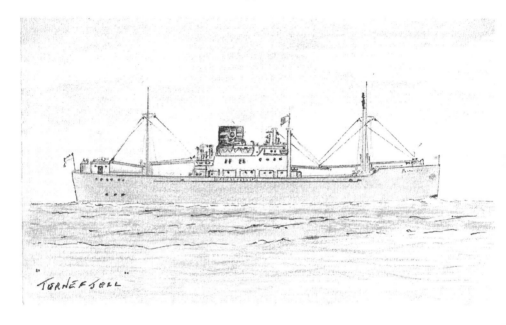

Ternefjell, by James Riach

London smog was not a pleasant phenomenon. Fortunately, it is no longer a problem. The Clean Air Act, which prohibited the outpouring of smoke from factories and homes, put an end to the appalling atmosphere, which was usually trapped beneath a cloud layer and was visible as a brown, all-enveloping blanket of germ-laden poison.

The last of the London smog is etched indelibly on my memory for two reasons. The first, and probably the most important, is that it witnessed the birth of my daughter, Aileen, who was born on 2 December 1962 and spent the first week of her life under a muslin cloth. The second is that it was the cause of one of my more exacting acts of pilotage.

In November 1961, I had been licensed as a junior, under draft, Trinity House pilot, stationed at Gravesend and operating in the London Channel district. I had satisfied my peers as to my fitness to conduct ships safely and speedily through the tortuous channels of the Thames estuary, providing no vessel was more than 14' draught of water.

Fog had been evident for at least 24 hours before I reported for duty on 1 December 1962. It was patchy and fickle, but weather conditions were perfect for a prolonged period of heavy and persistent smog.

The pilot station on the Royal Terrace Pier was quieter than usual and ships had been working their way down the river as and when they could. Many sailings had already been cancelled and scores of ships had prudently anchored, either in the upper reaches of the river or below Gravesend.

By the time I had arrived at the head of the evening roster list, it was obvious that weather conditions were worsening and visibility was becoming very bad indeed. By 0100 on 2 December, we knew three ships were in the offing. A large German cargo liner and a German coaster were already leveling down in Tilbury lock. Both ships could either be held in the lock, blocking all traffic in and out of Tilbury dock or, if there was a blink of clearance in the weather, they could be hauled out into the river. The decision to proceed or anchor was then the responsibility of the people on board. The lock master in Tilbury would certainly be anxious to see them sail and have the lock cleared.

The third ship, the Norwegian *Ternefjell* – she had a 14 feet draught and a gross tonnage of 1,001 tons – was ready to drop off a river berth at the Imperial Paper Mills, a mile upriver from

Gravesend and almost within hailing distance of the demarcation line between the river and channel pilots' districts.

In the early 1960s, communications and operative matters surrounding pilotage in London were verging on the primitive as compared to today's high technological systems. Radar, the all-seeing eye, was neither universal nor sophisticated, and VHF radio had yet to become the revolutionary factor of later years.

On the pilot station, we had direct contact by telephone with the Port of London's operations room, housed in the adjoining building, and we were kept informed about ship movements and the general situation on the river. However, as soon as a ship cleared into the river, they were out of touch and, not infrequently, ships slipped through the net and appeared without warning. We often had to depend on ships' whistle signals when they required a pilot – one long blast followed by four short blasts. Pilots were therefore obliged to stand by, sometimes for long periods on the pilot station, and be prepared for any eventuality.

The mood in the pilot station was one of anxious anticipation. I was at the head of the roster list but could only handle two of the ships expected. The *Ternefjell* and the German coaster were within my limitations, but the German cargo liner was most definitely not. The tension was broken by one long and four short blasts of a ship's whistle, and we were in business.

The second pilot on the roster list was an old friend of mine, Captain James Reid, a man of outstanding ability and supreme confidence, and a pilot for whom I had the highest respect. The majority of the pilots on the station were of the opinion that the *Ternefjell* was signaling for a pilot and she would be wise to anchor at once. However, the ship had clearly indicated her requirement for a channel pilot and Jim Reid and I decided we should make every effort to comply.

The distance from the pilot station to the riverside, where the pilot boats lie, was very short, but our vision was so restricted that simply getting to the boat was an adventure. When we did embark, we found the skipper of the pilot boat was less than enthusiastic about our plans. He thought that to set off blindly looking for a ship in these conditions was not a good idea, but he really had little choice, and we cast off heading north by compass in the direction of the whistle signal. Jim Reid and I took up our station in the bow, directing the skipper with hand signals when necessary.

After a while, we began to think that the *Ternefjell* had anchored, then suddenly we were alongside the ship. We began to move the boat along, hoping to find the pilot ladder but, breaking away a foot or two from the ship's side, we lost her and once again were floundering around with nothing to guide us.

We decided that we had better steer south and pick up the riverbank. The skipper was full of misgivings and, although we could not even see the wheelhouse, we heard him muttering about anchoring and how crazy we were. A dim red light suddenly appeared and we found ourselves alongside a jetty well to the westward of the pilots' pier. Nothing daunted, we crept cautiously along the riverbank. We were back where we started with nothing accomplished and were about to give up when the *Ternefjell* whistle, which we had come to recognise, blasted out again very close to us.

'Let go,' yelled Jim Reid, and a moment later we had cleared the pier. This time we were alongside the ship in seconds and under the pilot ladder. I leapt for the bottom rung, clambered up, gained the deck and made my way to the bridge.

It was very dark. The captain and the river pilot were standing like statues in the centre of the wheelhouse, and the only light came from the radar screen and the compass. The river pilot greeted me and said, 'Good morning, James. We are heading 080, engines on slow ahead. We're a bit close to the south shore but there seems to be nothing moving ahead or astern of you for a mile or so. She's bound south and the captain is a little concerned about the weather!'

With that, he shook the captain's hand and was gone.

The captain, a tall, distinguished-looking man, was obviously worried. We shook hands and I gave my first orders after a glance at the radar screen: 'Full ahead, steer 075.'

I was very much aware of the tension on the bridge and of my own concern. One of the first tips I had learnt from wiser pilots' heads than my own was that, no matter what the circumstances, a good pilot must always be cool, calm and radiate confidence. So my first order, full ahead, was reduced to slow ahead as soon as the ship began to move.

The captain, standing at my elbow, said, 'Pilot, do you think it might be better to anchor? I am in no hurry, and the steward has made up a cabin for you.'

At this stage, we were slowly moving down Gravesend Reach. There were other ships, barges and tugs anchored on the south side, their bells ringing to indicate their position. I had a clear eye ahead but I knew that, once we rounded the first bend in the river – which was two miles ahead – and entered the Lower Hope Reach, I would be confronted by a number of coasters and colliers anchored in the middle of the river and they would present an almost impossible barrier even in clear weather.

The chief officer and the carpenter were on the fo'c'sle head and I did not think that, to proceed against the captain's wishes and into a potentially dangerous situation, was prudent.

I angled the ship south again and came to an anchor as far off the main channel as I dared. The bridge cleared, leaving the officer on watch, one sailor and myself. I was reluctant to give up, and paced the bridge, drinking coffee, whistling for a breeze, and wondering if I would be confined to the ship perhaps for days.

After about an hour, I became aware of a slight improvement in the visibility. I suspected that it might only be a temporary lift, but I was determined to take every opportunity to move on and, as I began to see the details of the barges close alongside, I decided to act.

The captain was not too keen to move and enthusiasm for any movement was not apparent. However, the anchor was weighed, but kept outside the pipe ready for instant use, and we began to feel our way down Gravesend Reach towards the sea.

Operating solely on radar, and with frequent alterations of course, we negotiated the hazards of the next couple of miles and rounded the Ovens buoy into the Lower Hope.

The visibility again deteriorated and excitement reached fever pitch. The Lower Hope was littered with coasters, colliers and various craft, haphazardly anchored and all but completely blocking the river. Between the blowing of our own fog signal and the clamouring of bells from anchored ships, the noise was considerable, and my concentration was total.

At the top of the Lower Hope, a large starboard alteration of course was necessary and, at last, I could see I had a reasonably clear eye ahead and felt myself relaxing a little. Up to this point, I had been focused on what was ahead, with little concern of the situation astern.

It was therefore something of a shock when I suddenly became aware, out of the corner of my eye, of the large bow of an overtaking ship within spitting distance of the port side of *Ternefjell*. I rang the engines to dead slow, tried to appear calm and assured the captain, who was lighting his pipe with shaky hands, that all would be well.

I learnt later that Jimmy Reid had managed to board the German freighter, which had cleared Tilbury lock and had, in the slight improvement of visibility, worked his way down Gravesend Reach and followed me through the maze of anchored ships in the Lower Hope and, with expectation of clear water ahead, had blown the overtaking signal on the ship's whistle and rang full ahead on his engines. The situation was exciting, but not particularly dangerous. We had full control of our ships and nothing was moving ahead of us on the radar screen.

Jim Reid rapidly drove away from us and, although visibility fluctuated from zero to a cable or so, I felt reasonably comfortable, and the tide was against us, which meant that we could manoeuvre the ship more efficiently than if we had a following tide, and we could also stop the ship quickly. The captain was still nervous, but I felt a good deal better. I knew that, if we could clear the Sea Reach and gain the outer estuary and open sea, the visibility would improve and we would be in a much more comfortable position to proceed with safety.

The deepwater channel ahead of us from Thameshaven to No 1 Sea Reach buoy, a distance of 15 miles, was cluttered with ships at anchor, but little was moving and I felt confident that the worst was over and that I could press on, especially with our shallow draught.

As we followed the buoys down the Sea Reach, picking our way between the anchored ships and frequently altering course to avoid fishing boats and small craft, I was aware that the atmosphere on the bridge was a little more relaxed. The captain lit his cold pipe, which he had been visibly biting, and the man at the wheel eased his grip on the spokes and shifted his position to a more relaxed stance.

Southend Anchorage was a mass of anchored ships, as was the general approach to the estuary. Vessels coming in from sea in clear weather were suddenly running into a wall of dense fog stretching right across the estuary, and the result was that more and more ships were dropping their anchors wherever they could. Jim Reid's ship had long since gone, bound north for the Sunk light vessel, and I knew I had seen the last of him.

My plan was to work my way round into the Ouse, pass the Red Sand towers and the Shivering Sand towers, and follow the Princes Channel to the NE Spit buoy, and then inside the Goodwin sands to the sea. Visibility was no better and the passage between the sandbanks had moments of extreme anxiety. However, halfway down the Princes Channel, we suddenly ran out of the fog. One moment, we could not see the foremast and, the next, I could see many miles ahead, to starboard, the coast of Kent to the North Foreland and, to port, the Tongue lightship and the open sea. The sense of relief on the bridge was palpable, and the tension dissipated. Looking astern, I could see the great wall of dense fog, brown, sinister and menacing, stretching all the way across the Thames estuary.

The remainder of the passage was comparatively trouble-free. The fog had gone and the route inside the Downs, round the South Foreland and close to Dover harbour, although very busy with shipping, was negotiated with the usual attendance to the sea rules of the road.

I landed on the pilot cutter at Dungeness, to be told that the *Terneffell* was the first ship for a very long time to land a pilot.

My journey back to Gravesend was not without incident, since the smog had disrupted all transport. Within hours, my daughter was born and I found myself plunged into a crisis of an entirely different kind. My daughter spent the first few days of her life under close supervision – anti-smog measures.

The London river came to a complete standstill and I was not called out again for five days, but that is another story.

9. *Venus*, 1968

There is a little-known timber yard and jetty 15 miles up the River Crouch in Essex going by the name of Creeksea. It is situated on Wallasea Island, on the opposite side of the river from Burnham-on-Crouch, the haunt of yachtsmen and marsh birds, a bleak place in winter but cheerful enough in the summer.

As a Trinity House channel pilot for the River Thames, my licence included this area and if a ship came down the river bound for Creeksea from London, then channel pilots provided the service.

On the face of it, this arrangement seems ludicrous. Ships bound from London to Creeksea were very rare indeed and pilots have been known to work through their whole career without ever seeing the place.

Pilots work on a roster system, so the destination and size of the ship has no bearing on selection. When a ship crosses the demarcation line between the river pilots and the channel pilots, the first pilot on turn has to take the ship.

As a candidate or training pilot, I had never had the opportunity, nor the desire, to go to Creeksea, but I had learned the theory and maritime geography of the area. I knew all about the tides, courses, distances, depths of water, dangers and so on, but my practical experience of the place was non-existent.

On 26 April, 1968, I was No 1 pilot on turn when the good ship *Venus* appeared around Tilbury Ness and demanded a pilot for Creeksea. I had known for some time that I was likely to be the pilot involved and was prepared for the eventuality. In fact, I was rather looking forward to the challenge and the adventure of the trip.

The *Venus* was a little Danish coaster. Her gross tonnage was 400 tons and her draught was 10′. She was no ocean greyhound, but rather a North Sea terrier wandering the coasts from the North Cape to Ushant and paying visits to obscure and little-known ports throughout Europe.

The Captain, who was leaning on the table in the wheelhouse, greeted me warmly. He was a big man, maybe 20 stones, but he was cheerful and, in spite of his bulk, was agile and, as it turned out, had an excellent sense of humour.

'Pilot,' he said, 'we are bound for Creeksea. I have never been to this place but you can tell me all about it.'

I had shipped at 1410 on a fine, calm spring afternoon and, in spite of my slight anxiety about the immediate future, I began to enjoy myself. At a top speed of 10 knots, we were not going anywhere fast but, in due course, we cleared the Lower Hope Reach and the Blyth sands, and I settled down for the long haul towards the warps and the Swin Channels. We passed Southend Pier and, from then on, we were in one of the loneliest parts of the Thames estuary. The Swin Channels were little used – the occasional yacht or Thames barge might be seen but, in our case, I was glad of the company of the sea birds.

The captain and I entertained each other with yarns mostly about sea affairs and we talked about Creeksea. Of course, I did not tell him that I had never been there before and, from my conversation, although I told no lies, I imagine he had the impression that I was very much at home on the River Crouch.

The other occupier of the wheelhouse, apart from the helmsman, was a large Alsatian dog. This animal was in no way aggressive, but I thought it prudent to keep a wary eye on him and, as time went by, I think he accepted me as one of the ship's company.

The Captain's wife, who had made the ship her home, kept us well supplied with coffee and sandwiches. The mate was the Captain's brother, so the ship was a family affair, not uncommon in European coasters.

The afternoon progressed very pleasantly and in due course we arrived at the Whitaker buoy, which marks the approach to the River Crouch. We had no direct communication with anyone ashore or afloat. However, the interested parties, including the ship's agents in London and Creeksea, the men who handled the mooring ropes, the officials at the timber wharf and the captain, were of the opinion that we would arrive at the river mouth too late to reach the berth with water enough to float the ship and still have sufficient daylight to enable us to navigate and manoeuvre safely. I had already discussed this with the captain and had more or less resigned myself to spending the night at anchor and making our way upriver to the jetty on the next day's flood tide.

We stemmed the tide and let go the anchor, with the Whitaker buoy close astern, at about 1730 on a glorious evening. I turned to the tidal prediction tables to check out the general situation. Low water at Creeksea was at 1915. I had not really considered going up the river on that night's flood tide but, after giving the possibility some thought, I began to change my mind. I figured out that if we got under way at 1930 and proceeded with care, we would have enough water to float the ship and to arrive off Creeksea with just sufficient daylight to make the berthing a practical proposition. Sunset was at 2016 and, with half an hour of twilight, I had until 2045 to complete the job.

Fog was forecast and, with no radar, it would be difficult to navigate the river with reduced visibility. I was also very aware that any delay would allow darkness to overtake us and, as there were no lights on the buoys marking the channel, I could be in trouble.

I checked and rechecked and decided it was worth a try. I could find a hole to anchor as a last resort, although such action was not to be recommended.

I went below and informed the captain. He asked me if the pubs would still be open when we berthed and his wife produced another cup of tea. I was confident, but I knew that there was little margin for error. At 1930 the anchor was clear of the water and we set off into the unknown.

The passage upriver began without any problems. I kept strictly to the channel, marked infrequently by small buoys. As the banks of the river closed in on either side and the Outer Crouch buoy fell astern, the sun began to sink ahead of us and the water began to shallow. There were a few patches to be avoided. I could tell that the ship was very near the bottom from the effect on the steering, but I kept control and we gradually worked our way upriver. With six miles to go to the berth, the sun set and I began to worry. One half of my mind was urging the ship on before darkness overcame us and the other half of my mind was urging the ship to hold back to allow the incoming tide to increase the depth of water. We passed the Inner Crouch buoy, and Burnham-on-Crouch came into view. Darkness was closing in fast, but I could still see enough to push on and I concentrated on getting the ship through the yacht moorings. Buoys, yachts and all manner of pleasure craft seemed to spring up all around us and it looked quite impossible to get through to the jetty.

As we approached Burnham and successfully passed over the Horse Shoal, which had given me cause for concern, I blew the ship's whistle – three long blasts – to warn everyone ashore that we had arrived. I learned later that the boatmen and other interested parties were in the pub, never thinking that we would attempt to come up on that night's tide and that, when the ship's whistle blasted through the open windows of the pub, they all tumbled out just in time to see us roaring past with the tide under us.

I needed men on the jetty to take our mooring lines and, fortunately for us, the boatmen reached the jetty in time. They told me afterwards that they could not believe their eyes when they saw the ship and were considerably worried for reasons which will become apparent.

The night closed in, but I had a sighting of the jetty and, to my horror, I saw that a ship was tied up at its centre, effectively blocking what should have been a clear berth for us either ahead or astern of her. The boatmen had not bothered to move the ship, thereby clearing a berth for us, because they had assumed that we would not come up the river until the next day's tide. There was, I knew, a berth inside the jetty but, as it turned out, it was occupied by lighters and out of the question for us.

Darkness was now total, and I was in trouble. We were driving up on a flood tide with the water becoming shallower, no berth where I could secure the ship, surrounded by yachts and their moorings, and hardly room to swing a cat.

The chart clearly stated that anchoring in the area we were presently occupying was prohibited because of cables and mooring lines lying on the riverbed, but I had to act fast and had few options. I ordered the man at the wheel to put the helm hard a starboard, the engines to full astern, and instructed the mate, who was already in the fore part of the ship, to let go the anchor on the bottom. The ship swung within a foot or two of the wharf and we were round stemming the tide and in a position to drop alongside had we a berth to drop into.

My heart was thumping in overtime. I could see figures on the jetty, so I yelled out to them that I intended to come in. Their response was negative. Jumping up and down, they shouted, 'You can't come in here, not enough room for you.'

My reply is unprintable, but they were left in no doubt that I was coming in ahead of the ship already berthed.

I angled the ship in to the jetty and, using the anchor and the engines, I slipped past the anchor cable of the ship already alongside and dropped on to the wharf. We managed to get a line ashore and, after a good deal of work, we ended up with the bow of the good ship *Venus* more than 60' beyond the end of the jetty and the forward mooring lines made fast to anything that could be found, trees, bushes and the like.

The captain was a very happy man. I am sure he thought that what had gone on was the normal procedure and I had carried out similar manoeuvres many times before. 'Now we go to the pub,' he said.

The rest of the evening was spent in a very convivial atmosphere and, next morning, I hitched a ride on a timber lorry to Southend and made my way home by train.

I have had several ships to Creeksea since the *Venus*.

The last ship was a 4,000-ton Romanian vessel with a draught of 17', but that too is another story.

10. *Black Watch,* 1970

Black Watch, by James Riach

On 13 May 1970, the Fred Olsen cruise liner *Black Watch* was scheduled to sail from London to Ijmuiden in Holland, where a dry dock was waiting for her arrival. Everything had been arranged and ordered for 7am on 14 May and any delay in the timetable was unthinkable.

The day had been fine and clear, but the wind direction, humidity and atmospheric pressure indicated sea fog was more than possible.

It was not until I was the No 1 upper draught pilot on the roster list and in line to attend the *Black Watch* that I began to think seriously about what was before me. She was a very fine ship. She was fast and comfortable, with excellent food and I was looking forward to a very pleasant, if routine, job to the NE Spit buoy, where I would hand over to the captain. My work would be over in three hours and I would be home before bedtime.

The *Black Watch* was an impressive sight as she rounded Tilbury Ness and approached the pilot station and I was more than happy to climb on board and take over from the river pilot. As soon as I stepped on to the bridge, I knew that I had an exceptional ship on my hands. The *Black Watch* was equipped with every navigational aid modern technology could provide.

The captain shook my hand and dropped the first bombshell.

'Pilot,' he said, 'I shall be obliged if you will take the ship across to Ijmuiden. We are booked for the dry dock in the morning and we must be there on time. I have had very little rest recently and your services would give me a much needed break.'

My immediate reaction was pleasure and anticipation. I was well rested and ready for what lay ahead and, after the initial shock – it was indeed rare for a ship of this class to request a pilot for the channel crossing – I found myself looking forward to a fast and trouble-free run across to Holland.

We set off down Gravesend Reach, turned into the Lower Hope Reach and reduced speed as we swept past the oil berths at Thames Haven and Shell Haven. Sea Reach was no problem in spite of a busy procession of inward-bound ships and in due course we found ourselves passing the Red Sand Towers and making for the Edinburgh Channels and the open sea.

The Edinburgh Channels consist of two buoyed channels together, very narrow and crooked, so that frequent alterations of course were necessary when navigating them. They constituted the main deep arteries to and from the English Channel and the Thames estuary and it was always prudent to avoid meeting another ship or overtaking another ship in their vicinity.

Passing the Shivering Sands Towers to starboard, I sighted ahead of me a ship outward bound and obviously much slower than we were. I studied her with interest but, at this point, without alarm.

I quickly identified her as one of the elderly British India cargo ships. A senior colleague of mine, Captain Les Baker, choice pilot for the British India Shipping Company, had boarded her some time before I boarded the *Black Watch* and was bound south down the channel.

In 1970, VHF radio was not in general use and communication between ships was difficult. I could see that no ship was inbound through the Edinburgh Channels and that Les Baker was about to use the North Edinburgh Channel. I could see no real difficulty, as, by my calculation, I should overtake the British India ship in a relatively safe position.

Suddenly, the ship ahead began to turn to starboard, and then altered course to port and swung across the channel. 'Not under command' signals appeared above her bridge and the situation became disturbing.

There are times in all pilots' lives when decisions have to be made very quickly indeed. Instruments are of no help and the eye and the brain have to be relied upon to provide the answers. This was one of those occasions.

I had several ways in which I could take action. They were all feasible and practical and did not involve any danger, although they were not manoeuvres I would take voluntarily.

I could attempt to reduce speed, I could turn and use the Black Deep to get to sea, I could reverse my course until the situation cleared or I could proceed at full speed and trust that I would find sufficient room to clear the British India ship before or after she blocked the channel ahead of me. I knew that, once committed, I would have to stick to my decision, and all the while we were fast approaching the point of no return.

The ship in front of us was athwart the channel, but still moving ahead. I could see that a gap was opening up astern of her and that I could slip through without involving myself in a complex or dangerous situation.

I ordered the helmsman to steer a course which would clear the British India ship and keep me in deep water. Suddenly, the British India ship started to turn to starboard, but I was committed, and whatever happened next, I would have to act according to the circumstances.

At full speed, the *Black Watch* swept past the stern of the other ship with feet to spare and, in a moment, we were clear. The relief on the bridge was palpable, even although the bridge staff were not fully aware of just how critical our position had been.

The *Black Watch* shot out of the Edinburgh Channel like a greyhound from a trap, and I turned her to port for the Tongue Light Vessel and the relative open waters of the Dover Straits. Within a short period of time, I began to realise that we were going to have other problems. Beyond the Tongue Light Vessel, I could see stretching from NE to SW a great wall of grey-brown substance which could only mean one thing – sea fog.

Clearing the Tongue Light Vessel and setting a course for Ijmuiden, I expressed my concern to the chief officer, who was the officer of the watch, and he immediately sprang into action.

The engine was put on standby, extra lookouts were posted, the bridge was manned by a special fog team and all our navigational equipment was checked and seen to be operating correctly. It was very obvious to me that this ship was well-used, equipped and, in all respects, perfectly able to deal with a situation of this kind.

From my point of view, I was faced with a dilemma. I could go by the book, which would mean delay and cost to the ship, or I could act as much as possible within the rules – make full use of modern technology, my own experience and the efficiency of the bridge watch – and make for Ijmuiden, with the deadline for the dry dock always in mind. I chose the latter course and the result was one of the most exciting and exhilarating passages I have ever made on a ship.

I was constantly fed information on the radar, targets which enabled me to make clear decisions on action to be taken and as we had to cross the shipping lanes and there were many ships in our vicinity, we were very busy and constantly altering course.

The passage through the North Sea was an extraordinary experience and time passed very quickly. As we approached the Dutch coast and dawn was breaking, the fog cleared as suddenly as we had run into it. The tension on the bridge dissipated and we picked up the Dutch pilot without further incident.

11. *Decca Surveyor, 1974*

Decca Surveyor by James Riach

For 36 hours, my ship had been dodging 60 miles west of the Shetland Islands on the edge of the continental shelf.

Dodging is a trawlerman's term for lying with the sea on the bow and with just enough headway to keep the ship in position and under control. We had been ordered to the area to carry out a survey of the seabed to continue the never-ending search for oil but, after dropping one marker buoy, the westerly gale roaring in from the Atlantic put a stop to all our efforts.

With a Force 9 wind and a weather forecast that indicated no improvement, I decided to run for Lerwick and the shelter of the Shetland hills.

Turning the ship round was a delicate and hair-raising manoeuvre but, carried out at precisely the correct moment, the ship responded gallantly and, with the great seas towering astern like angry mountains, the ship surfed and planed towards Sumburgh Head and Lerwick harbour.

We approached Lerwick in squalls and poor visibility. The cliffs looked forbidding and stark, but the sun came out with the pilot and transformed the scene as if by magic. We had been keeping a bleak and lonely station out in the Atlantic with gannets and fulmars for company, where great grey seas, spume and constant violent motion were normal; and now, suddenly, entering Lerwick Bay, it was as if a door had opened and we were steaming into an entirely different world.

We had been preceded and were followed by deep-sea trawlermen. The harbour was pulsating, ships on the move, gulls screaming, quays and jetties alive with people, flags fluttering, all enveloped in an atmosphere, a cloak of indefinible character, that I have found only in Lerwick and nowhere else in the world.

I had never been to Shetland and I have not been back since, but that first impression and the experiences that followed will never fade from my memory.

The following morning, I had arranged to visit my agents on ship's business. Their office was situated in the main street and, after breakfast, I set off. The great west wind was still sweeping clouds low across the hills behind the town and out across the bay, but blinks of sunlight were frequent, and the grey stone and granite buildings looked as though they had just been scrubbed down by the sea, rinsed by the rain and dried by the wind and sun. Grey stone towns with flagged streets are generally gloomy places – not so Lerwick. It was summer and the streets were busy with holidaymakers, as well as local people and fishermen. Gulls were everywhere, screaming and squabbling, their plumage pearl grey, jet black and almost dazzling white. They must be the best-fed and healthiest birds in the world.

The agent's office was a large sprawling room, dark solid wood, friendly faces, telephones and activity. The fishermen were in port and they brooked no delay. The storm may ease soon and they had no time to waste. I sat down and spent the next hour enthralled by the scene before me. The fishermen almost to a man were Icelanders or Scandinavians, with the stamp of the far northern waters on their faces. They wanted ice, radar repairs, medical attention, oil, stores, a thousand and one things done and they all had fish to land. They were nearly all big men with ice-blue eyes and their dress was unique. Three skippers were vying with each other for service, each declaring in a mixture of their own language and ours that their needs were paramount.

The man on the left, one of the old school and as tough as old boots, had obviously dressed for the occasion. He had donned a collar and tie and a fine dress suit jacket, but had apparently decided that that was as far as he could go. From the waist down, he was all fisherman – Fearnought Trousers and fishing boots. The second man must have started from the other end, his trousers and shoes were almost Savile Row, but his smock and kerchief around his neck were straight from the fishing grounds. The third skipper was the strangest of all – a much younger man than his colleagues, his outfit was 'new generation' – tight, flared trousers, short jeans jacket, and his shoulder-length hair lay along a shirt which would not have looked out of place in Carnaby Street.

Their common denominator was the sea. Each man still had the westerly wind in his hair, their faces were alive and they had that tough healthy appearance of men well used to hard work and fresh air, and their eyes were real sailors' eyes, as blue and clear as the waters upon which they earned their living and as keen as the wind from which they had sought shelter. These men impressed me and, wherever they are now, I hope they are enjoying fair weather and fish aplenty.

The Round Britain Yacht Race was in progress and many of the boats were sheltering at Lerwick, jostling together, their rigging tinkling like bells. They were eager to be away again but were wisely waiting for more clement weather.

During our run up to Lerwick, my ship had been involved in a rescue operation for one of their number, 30 miles out in the Atlantic, and although the crew was rescued, the yacht was 50 miles away from where she had overturned when she was spotted by an aircraft and eventually taken in tow by a fishing boat. Intrepid and courageous sailors, these yachtsmen. They have to be, in those dangerous northern waters.

I had fresh fish for supper that night – a gift from a Norwegian boat on the next berth. Perhaps the black guillemot fishing close to the ship was equally pleased with his meal. At any rate, he looked as if he had dined well.

Next day, I flew south to London, returning to tube trains and pasty faces, strikes and city streets. I had finished my tour as relief captain and never again did I have the opportunity to revisit Lerwick.

However, the impressions gained on that short visit are cherished in my memory, and Lerwick and the Shetland Islands, the seamen and the people of the town are, I believe, blessed with a quality which is all too rare in our so-called progressive, modern society.

12. *Ardshiel*, 1977

On 19 February 1977, the P&O super tanker *Ardshiel* was due to be sold to the Greeks. The ship had berthed at B jetty, Thames Haven to discharge what remained of her cargo and the intention was to bring her up the River Thames to berth on Tilbury repair jetty where the inspection and handover would take place.

I believe that, at that time, she would be the largest ship to be berthed at Tilbury and great care would have to be exercised. It was essential that her draught of 25 and a half feet would make the passage through the Lower Hope and Gravesend Reach, so when the discharge was completed the ship was left without ballast and stood towering above the skyline and dwarfing everything at Thames Haven.

As an Inner List Pilot, I was instructed to board the vessel and take her upriver. This act of pilotage presented no apparent problem. The distance was short and the time ought to be limited and I expected to be home for tea.

The *Ardshiel* was a formidable ship more than a thousand feet long with a 158 feet beam and a deadweight tonnage of 214,085 tons. When she was built in 1969, she was one of the largest ships afloat.

The tanker berths A and B at Thames Haven were suitable for VLCCs – very large crude carriers. They were equipped with an elaborate tower which enabled the access to the ship to be a practical and reasonably comfortable means of boarding. A series

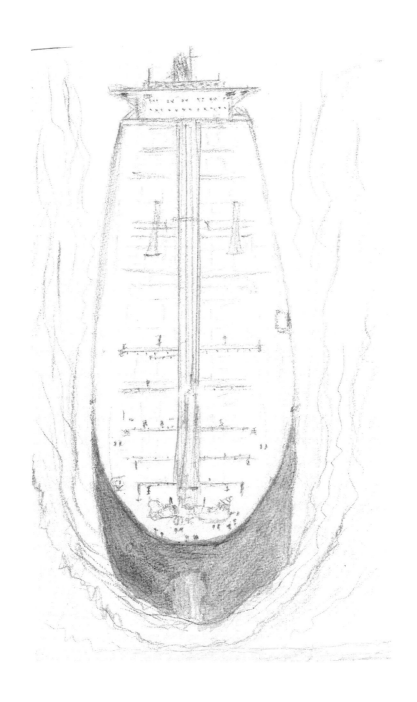

Ardshiel, by James Riach

of ladders within the tower led up to a gangway, which extended from the top of the structure to the main deck of the ship. But there was a drawback. The operation of rigging or dismantling the portable tower took considerable time and effort.

When we – I had a training pilot with me – arrived by taxi, we found the tower had already been completely removed, and boarding the ship meant a lengthy climb up a vertical pilot ladder. This presented no difficulty to us and we were soon on our way to the bridge deck.

We were ushered into the captain's cabin, and I was a little disconcerted to find a full-scale conference in progress. The captain gave us a warm reception, but I sensed that something was very wrong.

The captain was a small, bearded man with an air of authority about him and he quickly explained that there was a very serious problem. I was introduced to a representative of the workers' union, a member of the Ports Authority and the company's agent, and the difficulty became evident.

The main fuel discharge pipe was still connected to the ship by half a dozen bolts. The union of shore workers refused to allow the ship's crew to disconnect the pipe and no shore worker was prepared to climb the pilot ladder, but insisted that the gangway tower be reassembled. This operation would involve several hours and result in the ship missing the tide. The discussion went on and I pointed out that the union action would delay the ship by at least six hours, probably longer, and we would have to take the ship to anchor at Southend until the next tide.

The union representative was almost persuaded to allow the crew to dismantle the pipe but, at the last moment, refused to co-operate, so the talking went on, much to the amusement of the Greek captain who was on board and due to take over the ship at Tilbury.

Financial rewards were mentioned as an incentive to the shore gangs and the costs of delay were discussed. It had almost been decided that the gangway tower should be re-rigged and the ship anchored at Southend, when an officer reported that the wind was increasing and the weather was deteriorating rapidly.

The captain and I made our way to the bridge and together we examined the prospects. I obtained an up-to-date weather forecast by VHF radio and it was obvious to both him and me that to try to move the ship under these conditions would be folly.

In a way, we were both relieved because the cause of the delay was now due to the weather and out of our control. It was decided that the gangway tower would be re-rigged, the pipe disconnected by the shore gang and that I should remain on board in case of difficulties and take the ship upriver as soon as conditions allowed.

I spent the remainder of the day on board and, after an excellent dinner, I settled down in a luxurious cabin, where, to my astonishment, I found that the cabin walls were actually papered, a phenomenon entirely new to me.

The wind was gale force from the NE and, before I turned in, I had a look at the mooring lines. I was satisfied that they had been doubled, but a ship of this size, with such an enormous area affected by the wind, was vulnerable indeed and I was pleased that the officers were standing watch on the bridge throughout the night.

At about 0400, I was awakened by a cadet who informed me that all hands were being alerted to go to stations and that the ship was in danger of breaking free from her moorings. This was bad news. I lost no time in getting to the bridge and immediately called the duty tug to stand by and alerted the port authorities of the situation. One mooring line had already parted and the ship was several feet off the jetty. The remaining mooring lines were bar tight. I felt very uneasy.

We all remained at stations for the next four hours and, much to everyone's relief, the ship held fast and the wind began to ease.

After breakfast, weather conditions rapidly improved and, by lunchtime, we felt that a start could be made to get underway. By that time, the gangway tower had been reassembled, the discharge pipe had been disconnected and the tide was right. Our way was clear to haul off the jetty.

The *Ardshiel* was headed downriver and normally, with the tide flooding and the assistance of four tugs, we would clear the berth and turn to starboard until we were heading upriver. This manoeuvre was attempted but, even with the maximum effort of four tugs, it became clear that the high bridge structure was preventing the ship from turning through the wind. During the operation, the ship was bodily moving south towards Blyth Sands and was in danger of grounding.

A complete change of plans had to be ordered – and quickly. The tugs responded promptly, crossing the bow and stern and, with help from the engines, the ship began to turn slowly towards the jetty on the north side of the river, which allowed the wind to help rather than hinder us. By this time, we were close to the southern edge of the Channel and the propeller was churning up the mud and sand. It was a close-run thing and it seemed to take a very long time before the bow had cleared the jetty and we were heading upriver, still afloat and in control.

The remainder of the passage was without incident and, in due course, the ship was berthed at Tilbury. I never saw or heard of the *Ardshiel* again, but I have no doubt that she plied the seas of the world for a long period after the handover.

For my part, what was initially a comparatively short act of pilotage had turned out to be a complicated and dangerous operation, but one that was useful in the lessons learnt and the experience gained.

13. *World Premier*, 1978

When the VLCCs – very large crude carriers – appeared on the Thames, they presented a problem. The oil companies, particularly Shell, were concerned about the handling of these very large ships.

The solution they favoured was to have choice pilots appointed – men who were attached to the company – so that the ships would be in the hands of pilots with continual experience and not in the hands of the turn pilots, who may only be called upon to pilot these ships at irregular intervals.

The oil companies had a valid point, but choice piloting, which at one time was common with all the major shipping companies, had fallen out of favour and the pilots were not happy to see such an iniquitous system return. Common sense prevailed and, after negotiations, the matter was resolved amicably.

The pilots supplied the oil companies with an inner list of pilots recruited, voluntarily, from the middle-age group, thereby ensuring that the ships would be piloted by experienced men who were young enough to take the extra duty in their stride. The inner list was made up of volunteers, because it was thought that it was better to all to provide pilots who wanted to do the work, rather than pilots who were forced to do so. This system worked very well and, in 1978, I was in my second year as an inner list pilot.

On 12 May 1978, I was stood off to pilot the VLCC *World Premier* from Shell Haven B Jetty to Folkestone. It was going to be a long passage and a long night, but the weather was fair and I anticipated no serious problems.

The *World Premier* was 233,931 tons deadweight, with a draught of 38 feet and a speed of 14 knots. She was managed by Marine Navigation of London on behalf of Hong Kong owners and chartered to Shell. Her crew were all Chinese except the master who was British and had newly joined the ship at Shell Haven. She appeared to be well found.

We sailed at midnight without a hitch. After swinging the ship with our attendant tugs and manoeuvring the ship into position, we let go the tugs and began the long passage to sea.

It had been agreed that, before a pilot assumed responsibility for these ships and become inner list, he should accompany a colleague for a number of trips to familiarise himself thoroughly with the work. This was no reflection on the pilot's abilities, but rather to get him used to the sheer size of these vessels. On this occasion, a good friend of mine was with me, so, in fact, the ship was getting two pilots for the price of one.

We had sailed on top of the tide and were comfortable enough as we made our way from buoy to buoy in the Sea Reach. Our main concern was not the size of the ship, but her draught. We needed the flood tide's increased depth of water to keep the ship afloat. The depth of water under the ship was of prime concern.

Half a dozen channels thread their way between the Thames estuary sandbanks that are regularly used by ships bound to and from the sea, but for our ship, there was only one option – the Black Deep.

We were facing a long night. The distance involved was in excess of 100 miles. We cleared the Black Deep and sighted the Sunk Light Vessel guarding the approach to Harwich and the northern channels of the Thames. The Long Sand Head had to be rounded with care and, picking our way between the shallow areas, we almost reversed our course and turned south for the Kentish Knock and Goodwin Sands. The dawn began to lighten the eastern horizon, and nothing is more beautiful than the start of a new day on a fair May morning out beyond the restrictions of the land. Gradually, the details of the ship appeared stretching ahead of us, a vast expanse of steel deck and pipes and one or two tiny figures of crew members attending to their duties.

In the fullness of time, we raised and passed the North Goodwin Light Vessel and, with the notorious Goodwin Sands close to starboard, I began to think of reducing speed and preparing to disembark at Folkestone. VLCCs, with their great bulk and deep draught, require time and distance to control their forward motion.

The approaches to Dover are always an area where great care is necessary. Ferries frequently cross in and out of the port, all types of ships are moving up inside and outside Goodwin Sands and the fishing fleets are going about their business. At this time of year, yachts can also cause concern.

In my experience, many yachts appear to be unaware of the inherent dangers they face when they place themselves in hazardous situations. Many of them slavishly adhere to the rule that steam gives way to sail and, as long as they are sporting a scrap of sail, even if they have an engine operating at the same time, they assume they are at liberty to pursue the most extraordinary manoeuvres.

With Dover astern and the ship gradually slowing down, my colleague and I began to prepare to take our leave. On the bridge were the master, the officer of the watch and the North Sea pilot who was to take the ship on to Brixham. The crew were working on deck, just forward of the bridge, adjusting the hoist which would be used to lower us to the pilot boat about 100 feet below. I had talked to the Folkestone pilot station on VHF and passed my Estimated Time of Arrival and other relevant information, and all that remained was to position the ship off Folkestone pier so the boat could take us off safely.

In order to understand what happened next, some explanation is needed to show our method of disembarkation. VLCCs have a very high freeboard when in ballast and it is not a practical proposition to either board or leave by the conventional pilot ladder. It was therefore necessary to design a hoist, a type of lift where the pilot stands on a small ladder or platform and is mechanically transported up or down the ship side. Various designs were used and they were, by and large, perfectly safe.

The *World Premier* approached Folkestone on dead slow and, standing in the centre of the wheelhouse, I gave the necessary orders to place the ship in position for disembarkation. The fast pilot launch was approaching, so my colleague left the bridge and was duly lowered by the hoist to the waiting boat below.

I wished the captain a good voyage, informed him of the course to steer when I had left the ship and made my way down on deck to where an officer and a group of seamen were waiting to see me on my way. I positioned myself on the hoist and signalled that I was ready to be lowered to the waiting boat. A seaman released the brake on the winch and I began my descent. Seconds later, when I was about five or six feet below deck level, the ladder cockbilled. In other words, one end of the ladder tilted up so as to become almost vertical.

My reaction was to hang on for my life. I had, at this point, no idea that all that prevented me from plunging 100 feet to my death was an eye splice jammed in a sheave. My cap disappeared and my bag had fallen into the sea. I was in a very inelegant position, but I was still alive.

The following few moments were high drama indeed. I was aware that something was holding fast and I shouted to the deck crew to hoist me back on board. I was then able to turn and see what had actually happened, and my blood ran cold.

The wire upon which the entire apparatus depended had parted in the eye of the splice and had run out as far as the first sheave in the metal bar, which was secured to the short length of pilot ladder. And there it had lodged. Fortunately for me, the splice was too thick to go through the sheave. Had it done so, the wire would have run through both sheaves, and the ladder, the metal bar and myself would have plunged 100 feet to the pilot boat directly underneath.

Two serious flaws were very evident. Firstly, the hoist design was at fault. Two davits were positioned on deck at the ship's side. A steel wire was eye spliced into one davit, and the wire then ran through two sheaves in a metal plate to a block on the other davit, and then to a small winch on deck. Attached to the plate was a small length of pilot ladder. The pilot stood on the small ladder and was raised or lowered by a man operating the winch. The whole system depended on one wire with no safety back up. Secondly, the wire itself was old, rusty and ready to break.

I examined the wire and found that, although it looked well maintained and oiled on the outside, the inside of the wire had rusted and was weakened to a dangerous degree. I was shocked and very angry and returned to the bridge breathing fire and brimstone. The captain was visibly shaken and the North Sea pilot, who would have been the next man to use the hoist when leaving the ship at Brixham, was ashen.

I demanded an immediate start to be made to rig an alternative method of leaving the ship and the crew began the lengthy task of clearing the heavy accommodation ladder and swinging it over the side, where, in conjunction with a conventional pilot ladder, it would facilitate my departure.

While all this was going on, the ship was drifting, and I turned my attention to placing the ship in a position of safety. I spoke to the pilot launch and to Folkestone pilot station and made them aware of the situation, and the pilot boat streaked away into the harbour. Two hours later, I finally left the ship, much to the relief of Dover coast guards who were worried about our lack of mobility.

I was naturally very thankful for my escape, but was determined to do something positive to prevent the same thing happening again. I wrote to Trinity House, the Department of Trade and Industry, the Pilots Association and everyone else I could think of who could have influence in the matter. I was gratified that all the authorities treated my complaints very seriously indeed and the particular design of the hoist involved was declared illegal and no longer acceptable.

From that day until I retired, a close inspection of all disembarkation devices was high on my agenda of essential procedures before leaving a ship. I still shudder when I think of what could so easily have happened.

14. HMS *Onslaught*, 1981

On 15 July, 1981, one of Her Majesty's submarines sailed from London bound for Copenhagen to take part in NATO exercises.

HMS *Onslaught* was a fairly conventional non-nuclear submarine, 623 gross tons, with a draught of just over 19 feet and she required a pilot to the Sunk Light Ship. Warships are not obliged to take pilots, but many do, especially foreign warships, and I gathered that this particular ship had an unpleasant experience coming in to London without a pilot and had decided to take our advice on the way out.

I was the first pilot on turn, it was a fine summer afternoon and at 1515 I set out from the pier to board the submarine coming down Gravesend Reach. I was mildly curious, never having been on board a submarine, let alone piloted one, and I was quite looking forward to a new experience.

My first problem was the difficulty of actually gaining the deck of the ship. The outside hull of a vessel of this type is rounded and does not lend itself to the normal method of using a pilot ladder to climb on board from a pilot cutter.

VHF radio contact with the *Onslaught* made discussions about the difficulty an easy matter, and before I set off on the cutter, I knew what I had to do. Our pilot cutter was fitted with a platform well above deck level and, with the submarine stopped dead in the water, the cutter could come alongside the bow at right angles to the *Onslaught*'s fore and aft line and I could step on board over the stem.

The idea was fine in theory and was, in fact, a practical proposition, except that the distance from the cutter's platform to the *Onslaught* deck was rather higher than we had thought. However, the problem was overcome by two very large and brawny matelots, dressed in wetsuits, being stationed at the bow of the submarine for the prime purpose of manhandling me on board. This they did with expertise and care and with little affront to my dignity. Once on the deck, the sailors escorted me aft to the conning tower, and left me to introduce myself to the captain and take over from the river pilot.

The conning tower of an operational submarine is not a comfortable place to be. There is very little room and no space to move about. Immediately behind what little deck space was available, a gaping hole opened up, leading down via a vertical ladder to the operations room of the ship, and presenting a grave danger to the unwary.

The naval officers were, as I expected, welcoming and very polite. The captain was perched on a stool of some sort and the rest of us were crammed together with hardly room to move and, of course, exposed to all that the elements could throw at us. This was my first sub, but not my last, and I was interested in the systems of operations.

Of course, I was also concerned with the handling of the ship and quickly found that there was no problem there. All I had to do was give the orders and someone in the depths of the boat would hopefully carry them out.

The restricted space irritated me, but the other occupants of the conning tower were well accustomed to utilising every inch, and we even had a full and excellent meal brought up from somewhere down below.

The charts were in slots like the side pockets of a motor car, and there they remained. Over an intercom system from below, a voice kept up a constant stream of inaccurate information about where we were, distance of the next target, bearings of this and that – all in naval jargon. No notice was taken of this chatter until I pointed out that, if we were following the information supplied by the ghost voice from the depths, we would be in a dangerous situation, if not aground. The captain snapped upright and delivered a verbal reprimand to his men, which left me full of admiration for his vocabulary.

Our passage down the Sea Reach and through the Barrow Deep was uneventful and very pleasant. The weather was calm and we had no problems. In due course, I began to ruminate on taking my leave from the submarine and transferring to the pilot cutter.

I discussed the matter with the captain and the officers and we concluded that, to attempt an operation similar to my embarkation was not an option as, without the advantage of the raised deck of the Gravesend pilot cutter, the distance from the bow of the submarine to the small motor boat, which would be my mode of transport to the large sea pilot cutter, was just too great.

Fortunately, the sea was relatively calm and, after some thought, the captain came up with a possible solution. He was not prepared to over-carry me to Copenhagen as it was strictly against naval regulations to have a civilian on board the ship during NATO exercises; he was ready, if uneasy, about entering Harwich harbour to put me ashore. His plan was to flood the forward tanks and lower the fore end of the submarine so that I could practically step from the bow to the motor boat. I thought the idea feasible and told the Sunk pilot cutter our intentions on the VHF radio.

Unfortunately, the operation of flooding the forward tanks began too late and the time taken to lower the fore end of the boat was longer than we had anticipated, so when we arrived at my disembarkation point, we were in no position to facilitate my departure.

I began to resign myself to the extra time involved until the boat was in a semi-submerged state, when I suddenly had a brainwave. On either bow of the submarine were two steel plates in vertical positions, which were normally lowered when the boat was submerged and they looked like the fins of some enormous fish. The Captain assured me that he could lower the port fin in a few moments until it was parallel with the surface of the water and only a foot or two above the sea. This seemed to me the answer to the problem and the exercise was put into immediate effect. The pilot boat was informed and brought up to date.

A considerable clanking and creaking accompanied the lowering of the fin but, in a very short time, all was ready and as we were approaching the pilot cutter, I prepared to leave. The small motor boat with its expert crew had no difficulty in running in alongside the outer edge of the fin and, having wished the captain and his officers a good voyage, I was escorted down the casing forward towards the bow. Suddenly, as from nowhere, four fully-equipped frogmen appeared and, with admirable care, helped me along the fin and handed me into the boat. The whole procedure had gone without a hitch, and I stored the experience for future reference.

The last sighting I had of the *Onslaught* was of her sinister profile slipping away to the north, raising her fin like a last benediction before she disappeared from sight.

15. *Winston Churchill,* 1981

The Sail Training Association is a highly commendable organisation which had, until recently, two three-masted topsail schooners under their management. The aim of the Association was, and is, to man these ships with young men and women as crew members, not to train them to be seamen but to give these young people from all walks of life an opportunity to develop their characters and to show them that teamwork, discipline and a devotion to duty could be satisfying, helpful and even enjoyable.

Two-week voyages were normal and, during that period, the youngsters were given an experience of adventure and hard work, sometimes under difficult conditions, which allowed them a taste of an environment completely different from their normal lives and a story they could tell to their grandchildren as being one of the highlights of their lives.

The two schooners, the *Winston Churchill* and the *Malcolm Miller* were not yachts; they were sailing ships of considerable size and complexity. They had an auxiliary engine and a permanent staff of experienced officers, and were always kept to a high standard of efficiency and maintenance. The permanent crew consisted of a captain, a chief officer, a bosun, a cook, an engineer and sometimes a purser. Each of these professionals was highly qualified and skilled at his work.

The captain kept the 8–12 watch, the chief officer the 4–8 watch and, of course, both were responsible for many other factors and

The *Winston Churchill* at Chatham

duties with which they were thoroughly familiar. The need for one other officer was evident and so a volunteer, designated the navigator, was utilised to complete the deck department and to keep the 12–4 watch. The navigator had to be a man who had the necessary qualifications, was suitable for the job and was willing to spend two weeks or more in a new and challenging situation.

It was not always easy for the STA to find the right man for the position. Trinity House pilots were pretty well ideal. A number of the pilots were involved with the STA, both on the executive side and in supplying navigators and, when I was approached to volunteer to help, I agreed, partly through curiosity and partly because the cruise which interested me was to sail in a very familiar and much-loved locality – the Western Isles of Scotland.

I joined the *Winston Churchill* at Plymouth on a bright morning in May and was immediately plunged into a hectic period of familiarising myself with the ship and her equipment and getting to know my shipmates. It was all quite fascinating and I was intensely interested in watching and taking part in the methods used to prepare the ship for sea.

Late on the same day, thirty girls, aged between seventeen and twenty-two, were to join the ship and we were scheduled to sail the following morning. How does one take a complex sailing vessel out to sea at such short notice with completely inexperienced crew? I was soon to find out.

When the crew did arrive late in the afternoon, the mechanism of induction swung into action. The girls were pounced upon by the chief officer and the bosun who were experts in handling this initial stage and had done so many times.

The girls were confused, a little defiant and more than a little afraid. They obviously had no idea what was to become of them and the CO and the bosun gave them no quarter.

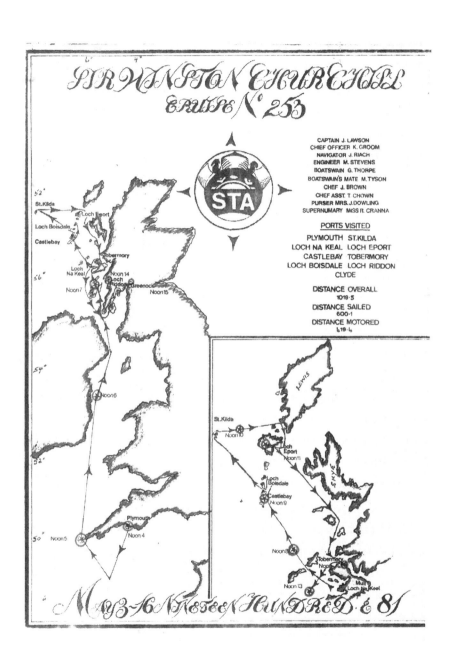

The *Winston Churchill*, Map of Cruise no. 253

They were split into two groups, the CO taking charge of one group and the bosun taking charge of the other. They were given a crash course of elementary seamanship and a tour of the ship, including instruction on the sail plan, rigging and the ropes involved in handling the sails. They were shouted at and left in no doubt as to who were the men in control but, most of all, the CO and the bosun impressed upon them the essential rule that, when they were ordered to heave on a rope or anything else, they would respond immediately and without question.

Finally, they were ordered to climb the rigging, over the mast and down the other side. White-faced and near to tears, the girls were then mustered and, rather reluctantly, the CO and the bosun gave them a few words of encouragement. They then had a very substantial meal and were bedded down for the night.

The following morning, everyone was soon aware that the coming weeks were going to be hard and tough, but full of interest. The ship was prepared for sea and, after taking on board a group of VIPs, including the mayor of Plymouth, we let go the mooring lines and headed out into Plymouth Sound.

We spent most of the day setting and taking in sail and tacking from side to side, which not only gave the girls much needed practise, mainly in hauling ropes and obeying orders, but it allowed the rest of the crew to settle into their duties. We landed our guests ashore in the afternoon after a buffet lunch and much wine, and at last put out to sea directly into a full south-westerly gale.

The girls had been divided into three watches: 8–12, 12–4 and 4–8, and each watch was in the charge of a watch leader. The watch leaders were young men or women who previously had sailed in the schooners and were seasoned and practical deep-sea yachtspeople. They were in control of the watches on deck under the overall charge of the officer of the watch and this system worked very well indeed.

Beating our way to the south-west in such conditions was not the ideal way to introduce our novice crew to life at sea. It was uncomfortable to say the least and, for the girls, it must have been sheer hell. They were allowed no respite and no sympathy. Watches had to be kept and most of the girls were seasick and miserable. None of them had suffered in this way before, and probably they would have jumped over the side if they had not been too afraid of us to do so.

At 0400 hrs, after a violently stormy night, I handed over the watch to the chief officer and made my way below. I had to descend a vertical ladder to an alleyway to reach my cabin, and I noticed that I was being followed by the diminutive figure of a very unhappy girl. She was clad from top to toe in oilskins, with only a wisp of hair and two eyes peeping out of the protective clothing.

She was wet, seasick and thoroughly miserable. She stood holding on to anything within reach and said, 'Are you the coxswain, sir? I've been told to call the coxswain.'

I really felt sorry for her. She had probably never been awake and expected to work at that hour in the morning in her whole life. To her, we must have appeared to be strange and threatening monsters. I said, 'OK, young lady. You have probably been ordered to call the cook. However, what you must do is to go back up the ladder and ask the Chief Officer to repeat his instructions and carry them out to the letter.'

'Yes, sir,' she replied.

The poor wee girl laboured up the ladder until only her sea boots were visible, and I heard her shout, 'What did you tell me to do, sir, because there is nobody down there but a grey-haired old gentleman.'

I considered myself at that time to be in my prime, without a grey hair on my head, and the girl's observation did nothing for my ego. I related the story to my shipmates at breakfast next morning, much to their amusement, and later attempted to identify the culprit. Nobody would own up and I never did discover who had cut me down to size on that stormy night in the English Channel.

We spent all night labouring, tacking and fighting to cross the channel. It was usual on each voyage, if possible, to take the ship to a foreign port, if only for a short visit. Our destination was to have been Douarnenez in France at the northern edge of the Bay of Biscay. The weather showed no sign of improvement and reluctantly we gave up and altered course to head up the Irish Sea. Time was not on our side and we simply had to get to our main cruising ground in the Western Isles of Scotland.

With the wind on the quarter, and still blowing very hard, we made excellent progress and were soon well on our way to the north.

The crew were not happy. They were still seasick and now they were also exhausted. Most of them had had little sleep since we sailed, and most of them had eaten very little. I began to worry about their condition and expressed my concern to the captain.

I told him the girls were in a bad way. 'Perhaps,' I said, 'it would be a good idea to call in at Douglas on the Isle of Man, or Belfast or even Campbeltown and give them a break?'

'Never!' said the captain. 'I do not intend to call anywhere where these girls have access to a train, bus, car or any other means of transport, for obvious reasons.'

I had to admit he had a point, but my sympathy remained with the girls. In the early hours of the following morning, one of the girls in the captain's watch had had enough. She downed tools, turned aft, and told the captain what she thought of him and his ship. After the initial shock, the captain roared out a reply which was enough to stop the girl in her tracks, and ordered her to report to his cabin at 1000 hrs.

During breakfast, the captain and I discussed the issue. 'That girl showed spirit,' he said. 'That sort of behaviour is not serious and must be dealt with delicately. I'll put the fear of God into her this morning.'

I couldn't wait to hear the outcome after the interview.

'Anticlimax,' said the captain. 'She burst into tears and apologised profusely, blaming the pressure of the moment for her action. I forgave her of course, but left her in no doubt that mutiny would be dealt with severely.'

I think every girl on the ship thought that the captain and the officers, including myself, were completely mad, devoid of any human qualities and would have happily cut our throats to get off and go home.

The wind began to ease off as we gained northing, but it was followed by torrential rain. We sailed on in decreasing visibility and crept into Loch na Keal on the Isle of Mull, where we dropped anchor for the night.

The relief felt by the crew was almost palpable. The girls began to blossom like spring flowers. No longer did they have to contend with a heaving deck, spume and spray and constant exhaustion. They were dry and, best of all, were rapidly recovering from seasickness. One of the first results of their freedom was to wash their hair and our fresh water tank level became alarmingly low very rapidly.

The captain called a conference of the afterguard. It was decided that we would heave up the anchor at dawn and sail through the islands and out to Castlebay on Barra, where we could replenish our fresh water supplies. We also discussed our bond. Had we been able to berth in France, the customs seal on the bond locker could have been broken and access to duty-free drinks would have been possible. However, even although we had been close to the French port, we had been unable to enter, and the legality of breaking the customs seal and utilising the contents of the bond was in doubt.

The captain gave us two options. We could leave the seal intact and become a dry ship, or we could break the seal and enjoy the contents, but be prepared to talk ourselves out of any problems and possibly have to pay full prices for our drinks. The decision was unanimous – and the bond was opened forthwith.

Under normal conditions, the charts are cleaned off after use, but I decided that I had better keep our chart of the English Channel intact for future reference in case of difficulties in dealing with HM Customs.

Next day, the weather was still unpleasant but at least the wind had eased to a fresh breeze and we set our course through the islands, passing Ulva and the Treshnish Isles and tracing our way through the Gunna Sound between Coll and Tiree, and out into The Minch.

Navigation in the Western Isles is a tricky business, especially for a ship with a deep keel. We were drawing 17 feet of water or more, so absolute concentration was required. We had come north through the Sound of Jura, passed Iona and skirted Staffa, before feeling our way into Loch na Keal. The trip from there to The Minch was littered with dangers. We had, of course, satellite navigation radar, echo sounder and Decca Navigator. However, great care was essential and human reactions to circumstances are always the ultimate safeguard. My training as a pilot taught me to keep away from danger as far as possible, and I like to think that I was a restraining influence on the captain, who was somewhat less cautious than I was and inclined to be rather adventurous at times.

I would have preferred a more detailed chart of Barra. None of us had been before and as visibility gradually became worse – drizzle and sea mist – we concentrated on finding the Fairway buoy and entering Castlebay.

In the very excellent book giving sailing directions for the Western Isles known as *The Pilot*, it is pointed out that the approach to Castlebay was marked by a pair of leading marks on a hillside and that by keeping them in line, a ship could approach and enter the bay in safety. However, we did not need to follow this to the letter as I had charted our approach and was quite confident that, by running fairly close to a ruined tower on the rocks and exercising caution, we would have no difficulty in gaining our anchorage in the bay. In the event, we did pick up the leading marks but they were not much help. They were difficult to see, blending in with the hillside a little too well. We let go our anchor close to Kisimul Castle and dropped our boat alongside.

Our main concern now was to top up with fresh water so to find when and if we could come in alongside the pier, I set off in our boat to locate the pier master. He was a tall, burly man, dark, middle-aged and very helpful.

'Are you off that schooner, is it?' he asked. 'They said you have 30 girls on board and all the young boys here are waiting for them with open arms, so you can come alongside anytime.'

'Right,' I said. 'Now what about fresh water, and what is your depth of water at the pier? We are at least 16 feet draught and maybe more.'

'Och,' said the pier master. 'We have plenty of fresh water coming from that wee hose there, and if you want any more, the good Lord will give you as much as you require by opening the skies. As far as the depth of water at the pier goes, well, we have enough water there to float the *Queen Mary*.'

I was a little sceptical, so I sounded the depth myself. And, of course, the pier master was right.

'OK,' I said. 'Now, according to *The Pilot*, there are lights on your leading marks on the hill yonder and, as we may be sailing at night, can you confirm this information?'

'Yes, yes,' said the pier master, with an expression on his face that was hard to fathom. 'Sometimes there is a light on one, and sometimes there is a light on the other and, now and again, there are no lights on either.'

Armed with this intelligence, I returned to the ship and informed the captain. We picked up the anchor and dropped alongside the pier, which, by this time, was crowded with the expectant youths of Castlebay and what appeared to be a fair number of the good citizens of the town.

Our stay at Castlebay, short as it was, was most agreeable. We spent the night alongside, which was a great thrill for the girls. They had been cooped up for five days and some of the older girls decided to sample the delights of the local hotel and bar. The captain invited me to go ashore for an hour or two, not to spy on the crew but just to make sure that they were all right, and to enjoy a drink at the hotel at the same time.

We arrived to find a line of girls at the bar, their glasses were full and they seemed to be enjoying themselves.

'What are you drinking, Mary?' I enquired of the nearest young lady. 'Well,' she said, 'You wouldn't take us to France, so we decided to have a Pernod. The barman spent half an hour looking for it and finally produced half a bottle, sir.'

Naturally enough, the girls were using the public telephone to call their homes and, on enquiring as to their success, one girl said to me, 'Well, sir, I talked to my father and he was very relieved to hear that I was fine and dandy. He said, "What port in France are you speaking from, daughter?" and when I replied I was not in France, he said, "Then where on earth are you?" "I'm in the Outer Hebrides of Scotland," I said. "The navigator took the wrong turning at the end of the English Channel."'

The captain and I returned to the ship a little wiser and a little more aware that our crew were beginning to show the signs of enjoying themselves and were beginning to respond with a touch of humour.

Next morning we were faced with a small crisis. One of the girls had damaged her wrist and, after visiting a local doctor, it was decided to make sure that nothing was seriously broken, we should sail at once for Lochboisdale where proper X-ray facilities would be available. Our friend, the pier master, waved us off and I determined to go out with the leading marks in line astern. We were motoring and not under sail and were soon making our way according to the book. All went well until I noticed we were uncomfortably close to rocks on the starboard side. The book was abandoned and we brought the ship out without reference to leading marks or anything other than common sense and my trusty chart.

The weather had not improved, but we got to Lochboisdale and tied up without incident. An ambulance was waiting and our patient was whisked away. We decided to wait on as long as possible in the hope that, if there was no serious injury, the girl could rejoin and we could proceed. The tide was ebbing fast and I worked out that we had about two hours before we would have to leave the pier and anchor off. I put on the echo sounder and kept a close eye on the reading. It would have been a disaster if the ship had grounded. Ten minutes before the deadline, the ambulance roared on to the pier and a very happy girl reported on board with nothing more than a sprained wrist. We hurriedly let go, swung the ship and headed for the open sea.

The captain and I had already discussed our next destination and, though he and I were the only enthusiasts, the captain's word was law, so we set our course for the Sound of Barra and the islands of St Kilda.

This, for me, would be the highlight of the cruise. I doubted if I would ever again have an opportunity to see these remote islands and, having read about them with great interest, I was delighted we were making the effort to visit them, even if the weather was atrocious and landing ashore would be impossible.

Navigating the Sound of Barra was really tricky and, under the engine alone, we threaded our way past rocks within spitting distance of the ship's side. We slipped past Eriskay, where the *SS Politician* had gone ashore with a part cargo of whisky, which was immortalised by Compton Mackenzie's book and the film, *Whisky Galore!*

We gained the open Atlantic with considerable relief.

For the next 50 miles, we had glorious sailing. The wind was gale force and we swept along under reduced canvas. We blew the mizzen sail out that night, which caused great excitement and made the crew feel that they were real sailors.

At 1600 hrs, I handed over the watch to the chief officer. We stood in the saloon hatchway directly facing the wheel and sheltered from the spume and spray. The helmslady was battling with the wheel and obviously not happy.

Unfortunately, she was one of the girls who never completely overcame seasickness and, though she never gave up and kept her watches, she was having a very rough time.

The CO and I had often discussed the psychology of handling a crew of very individual, inexperienced and modern young ladies. His philosophy, although sounding simple was, in fact, full of complexities. He maintained he must keep the pressure up at all times and believed that to show sympathy at the wrong time would be detrimental to the discipline and the smooth running of the ship. He and the bosun were very strict indeed. However, when a word of encouragement served its purpose, they would relent, but that was a rare occurrence.

We were being accompanied by seabirds of various species and, as we stood in the saloon hatchway before the wheel, a bird flying overhead unloaded at exactly the right moment, to mess

The *Winston Churchill* at sea

up both the wheel and the young lady who was steering. The CO immediately ordered the lookout to get a wet cloth from the galley and I thought that at last some sympathy was going to be shown to the unfortunate girl and that the cloth was for cleaning her up.

The lookout returned with the cloth and the CO ordered her to wash down the wheel. The poor girl was completely ignored. She was a picture of misery and my heart went out to her. However, perhaps if the CO had acted differently, she may have broken down completely – who knows? The same young lady will feature later, but, for the moment, her world was black indeed.

St Kilda was breathtaking. The sea cliffs are the highest in the UK and, seen from the deck of a small sailing ship, they were forbidding and very impressive.

We sailed across the entrance to Village Bay and scanned the area with our binoculars. A landing was impossible, so we eventually turned, left the islands astern and set our course for the Sound of Harris.

The weather began to improve steadily from the west and patches of blue sky appeared. The captain was for pressing on but, as we were in danger of losing daylight at the eastern end of the Sound of Harris, I persuaded him a better option would be to anchor and tackle the Sound at dawn. In the event, we dropped the anchor off Northtown and spent a quiet and restful night.

Next day dawned bright and fresh – perfect weather – warm and sunny. We picked up the anchor and, using the engines alone, had an exciting passage through the Sound of Harris. Rocks appeared to spring up from the sea bottom all around the ship. We were guided by leading marks and buoys, but I must admit to a great sense of relief when the tension slackened and we were safely through into The Minch.

We swung to starboard, slipped past Weaver's Point and Lochmaddy, and squared up to enter Loch Eport, a very long and narrow sea loch which penetrates the very heart of North Uist. The entire area is remote and sparsely populated and, as we moved up to an anchorage at the very head of the loch, we were all impressed with the scenery. It seemed as if there was no one else on the planet and, to everyone's delight, the sky was blue, the sun was warm and the wind a balmy breeze. Our boats were put in the water and the girls set out to explore.

I decided it was time that I climbed the rigging and so, armed with my camera, I announced my intention. My watch of girls immediately expressed concern and, after a huddled conference, I was told that three minders would accompany me – one girl on each side and another below. It was all unnecessary but very flattering. The view from the top was breathtaking, and I secured several good photographs of the ship from a considerable height.

The captain and the CO were fully occupied repairing the mizzen sail, but our time was limited and, rather reluctantly, we eventually had to make our way back out to sea, taking with us memories of a very peaceful and delightful stay in a seldom visited, but very beautiful, anchorage.

We had a splendid sail across The Minch to the coast of Skye and raced south through Canna Sound and swept past Ardnamurchan Point, the most western point on mainland Britain. We entered Tobermory bay in great style and came to an anchor close to the town.

By this time, our crew were fighting fit and had adapted to the routine of the ship. Seasickness had been overcome by most of the girls and they were actually enjoying themselves. The food on board was excellent. The weather was glorious and the girls were at last recognising the value of what they were doing. And, best of all, we, in the afterguard, were no longer the cruel taskmasters, but were beginning to be regarded with more heroic properties.

The girls thoroughly enjoyed Tobermory. They manned the boats, they explored the town, did their shopping and were ogled by the local lads. We spent a full day and night in the bay. I had been partly brought up in Tobermory, and the Mishnish Hotel on the seafront was well known to me. The afterguard had dinner at the hotel under the hospitable wing of Bobby Macleod, the owner. The meal was agreeably formal and was enjoyed by all of us.

Next morning, with the mizzen sail overhauled and repaired, we cleared the bay bound for the Clyde. I had suggested proceeding south through the Sound of Mull, but the captain decided otherwise and we skirted the north coast of the island and sailed through the north passage and around the Mull of Kintyre, moving up between Arran and the mainland, approached the Kyles of Bute and anchored at the head of Loch Riddon.

The north end of Loch Riddon is very shallow and the captain was keen to anchor as close to the shallows as he dared. We approached cautiously, with the captain on deck above the saloon, conning the ship and myself below in the saloon where all the navigating instruments were located. My job was to furnish the captain with all the relative information, gleaned from the radar mostly, by shouting up through the skylight. It worked well enough, but I found myself willing the captain to drop the anchor as we came closer and closer to the shallows.

We eventually did drop the anchor but, by that time, I was more than a little concerned and was greatly relieved when I heard him give the order to let go. As soon as we had settled down, the captain called a conference of the officers. To my surprise and amazement, he announced that there was going to be a concert that night.

Apparently, this was the normal procedure on the last night of each cruise, and all of us would contribute and perform some act of entertainment. The permanent members of the afterguard were well prepared for the event. They had perfected their acts through many performances and knew exactly what they were going to do.

To my consternation, I was ordered to arrange the party immediately and that the festivities would begin with a beauty contest at 1800 hrs. This contest was also a regular feature of a girls' crew cruise, and the CO and the bosun were the organisers. The CO was the chairman of the judges who were the purser, the bosun and myself.

Before the event, the CO gathered the judges together and put a proposition to us that was not to be questioned. The girl who had been on the wheel on the way out to St Kilda and had been so misused by a passing seabird, had never really recovered from seasickness and her cruise had not been entirely happy. The CO informed us that she would be the winner of the beauty contest. No one objected and the CO went up a notch in my estimation. That girl had had a very tough time and at least she would have a positive memory of the beauty contest to look back upon. The girls entered completely into the spirit of the competition and turned up in various rig-outs. The St Kilda girl was awarded first prize and everyone was happy.

The party that night was an outstanding success. To my astonishment, I found considerable talent among the girls. I winkled out two guitar players complete with instruments, a couple of excellent songsters, poetry readers and even one girl who was a very good comedienne and mimic. The CO and the bosun appeared completely dressed as women, with mop wigs and all the trimmings. They had obviously rehearsed and perfected their turn over many cruises and sang 'Sisters' to great applause. The captain appeared with a little sailor's hat on, clutching a ukulele and destroyed his past image of Captain Bligh by singing a couple of comic songs. I even contributed to the event myself by giving a rendering of one or two sea shanties and the girls roared out the choruses with evident pleasure.

Next morning at dawn, we were under way and sailed through the Kyles of Bute with great style. We had a glorious day's sailing in the Firth of Clyde and Loch Long and, in the late afternoon, slipped alongside the jetty at Greenock, and the 253rd cruise of the *Winston Churchill* was over.

I had one last duty to perform – a task which had dogged me and caused me concern ever since sailing from Plymouth. Our bonded stores of duty-free drinks had been utilised by the afterguard throughout the cruise and, although I had made every effort to find a customs officer to clear the ship, I had been unsuccessful in our ports of call. Greenock, however, was our final port of entry and I was delegated the delicate responsibility of entertaining the customs officer who came on board, the unanimous feeling among the afterguard being that I could speak the language and that I was best fitted to deal with the situation.

I produced the charts, showing him our planned route and the course of our voyage, and the logbook, which I had kept carefully, and after a pleasant chat with a very helpful customs officer, the ship was entered in without any problem concerning our consumption of duty-free stores. I didn't exactly explain the closest details of the trip and he didn't enquire too deeply either, so we were accommodated without difficulty. My reputation and general standing shot up with the afterguard and I was the hero of the hour – a fitting end to an eventful cruise, and I was very pleased with myself.

That night, a party was held at the Greenock Yacht Club, and the girls left the ship the following morning. The departure of the crew turned out to be quite emotional. The 30 young girls who had come on board at Plymouth, tearful, anxious and unaware of what was to happen in the next two weeks, left the *Winston Churchill* with tears of regret and promises to return. We, in the afterguard, had started out as devils and ended up as heroes. One thing is certain, these girls will never forget their introduction to the sea and, as most of them admitted, they had gained many valuable lessons which would be of enormous benefit to them in the future.

On the deck of the *Winston Churchill*

A new crew was due to arrive after the girls' departure, a mixture of public school and remand home boys. The CO, who applied his own brand of psychiatry to the situation, had arranged to have the remand home boys on board 24 hours before the public school boys so that the balance of inferiority and superiority could be contained. The CO told me later that there had been no trouble on the cruise and that, with a little help from the bosun and himself, the crew had behaved well as a team and had cemented friendships, which could only be of help to those youngsters.

The sequel to the story takes place in the Captain's Room at Lloyds of London. Every January, the participants in the previous summer's cruises are invited to a cocktail party and I made a point of attending.

Our girls were there in force, but I scarcely recognised the pretty, well-groomed and sophisticated young ladies who appeared, in contrast to the tough seafarers I had last seen at Greenock. To a girl, they were delighted to see us again and they all said that they would never forget their voyage to the Hebrides and the hard and happy times on board ship.

They all considered that the cruise had a marked effect on their lives and some of them were determined to return for another cruise and eventually to become watch officers. I believe we all learnt a good deal from our period together. The girls learned how to work as a team, as well as the advantage of doing something completely outside their normal day-to-day activities and doing it well. They learned to cope with difficult and sometimes potentially dangerous situations. They learnt tolerance to others and that each human being is an individual and capable of much more than she expects of herself. This sort of experience can only be of benefit to youngsters of all types, and encouragement should come from all sectors of society to support and foster the Sail Training Association's activities.

16. *Dashiqiao*, 1983

At 1615 on 20 January 1983, I boarded the Republic of China cargo vessel, *Dashiqiao*, to conduct her from Gravesend through the channels of the Thames estuary to an anchorage at Margate Roads. The ship was bound for Rotterdam, but had been instructed to anchor off the North Foreland until it was suitable for her to proceed. The reason for the delay was not clear to me.

The *Dashiqiao* was a medium-sized typical cargo ship. She was 6,936 tons gross, her draught was 22′4′ and she was capable of the not inconsiderable speed of 18 knots. The Captain's name was, as far as I was able to decipher, Chenzhi.

I was not displeased with this allocation of duty. A fast ship to Margate was one of the better jobs for pilots; it was comparatively short and I had every prospect of being home by midnight. The weather was fair, the wind was NW 5, with good visibility, and I was looking forward to a pleasant, if routine, act of pilotage.

As was usual on Chinese ships, I expected they would speak little English, but I did expect that the sea terms necessary for the operation of the ship would be readily understood and acted upon and that my orders would be obeyed with alacrity. To my surprise, I found that both the captain and the chief officer spoke English, even if their interpretation of the language, their expressions and accents were entirely their own. I determined to seize this opportunity to discuss matters with them which, in normal circumstances, would be practically impossible.

The passage to the anchorage was without incident. I was able to utilise full speed after I had negotiated the river through the Lower Hope Reach, carefully slowed down passing the oil jetties at Thameshaven, Shellhaven and Canvey Island, and put the ship on course for No 7 Sea Reach buoy. We swept past the Oaze banks in great style and with the ebb tide under us, the Red Sand and Shivering Sand Towers dropped astern. We had no problems from other ships or anything else.

Because of our draught, I decided that there would be insufficient water for us in the Princes Channel, so I conned the ship up to and through the North Edinburgh Channel. Once clear of the buoys at the southern end of the channel, I swung the ship to starboard and set a course for the NE Spit buoy and the anchorage.

At 1915, after the usual manoeuvres, I stemmed the tide and let go the anchor. The NE Spit buoy was well outside, and we were less than a mile from the North Foreland. I was well pleased with our position.

Up to this point, there had been little chance for general conversation. I had to concentrate on taking the ship through the channels and, with our considerable speed, I had to give all my attention to the safety of the ship. However, with the anchor securely on the seabed, the atmosphere on the bridge relaxed and my mind turned to other matters.

I began by probing gently into the philosophy and the differences between the British and the Chinese cultures. The captain, smiling, as was usual with the Chinese, was perfectly open and forthcoming.

He first insisted that we all have beer, which was produced as if by magic by a steward who had been hovering in the background.

'Pilot,' said the captain, 'you are a man from Scotland, a Scottish man.'

'Indeed I am,' I replied. 'How very perceptive of you, Captain. How did you know?'

The captain and the chief officer exchanged glances and then the captain stated, 'I can tell from your accent.'

'Of course,' I said. 'So your ship must have visited a Scottish port at some time, perhaps Glasgow?'

'No. I have never been to Scotland.'

The conversation then became a little more guarded but eventually I discovered that there had been a Scottish instructor in the school of navigation in China and his accent had alerted the captain to my nationality. Since then, I have often wondered how on earth a fellow Scot could have secured such a position.

I had, by this time, contacted the Margate pilot cutter on the VHF radio and was informed by the skipper that they were presently shipping two pilots on board inward-bound ships at the Tongue Light Vessel and that they would come and take me out on their way back to Margate.

I had another half hour on board the *Dashiqiao* before disembarking and, as I was being treated to such splendid hospitality, the prospect of further delay was not inconvenient.

The captain, chief officer and I drank some more beer, and suddenly the chief officer, who had been comparatively quiet up to this point, said, 'Pilot. We, in China, are very happy with the work of your country's bard.'

I was taken a little aback by this announcement, but soon rallied and said, 'You must mean Robert Burns.'

'Yes, yes, Mr Pilot, Robert Burns. We know him very well. You know, "Should auldy langy syney be forgotty, and never come to mind." '

I replied that I was familiar with the work of Burns but that I was no expert. The captain then demanded that I write down the words of the song forthwith, and I was escorted into the chartroom and supplied with pen and paper. I began to dredge my memory.

Fortunately, I did remember the first two or three verses and so, to the delight of my expectant spectators, I wrote them carefully down on the paper. The captain and chief officer were overjoyed. They jabbered away in Chinese, seized my paper, and there and then gave me a rendering of 'Auld Lang Syne', the like of which I have never heard before.

I was charmed, but I had to correct my two friends on the musical rendering of the song. Their handling of the melody was definitely not musical to my ear. I did my best and my friends were very eager to learn. My own singing voice was no threat to Pavarotti, but it served the purpose, and after another beer and much amusement, we considered that Robert Burns himself would have been proud of us.

It could have been a historical moment in the cementing of international relations had a video recorder been on hand to record it for posterity. I imagine the chorus comprising a normally dignified Trinity House Pilot and two Chinese ship's officers singing their version of 'Auld Lang Syne' must be unique and almost in defiance of imagination.

My pilot boat arrived alongside and I had to take leave of my fellow songsters. The captain and chief officer were effusive in their appreciation of my assistance, and to this day I never sing 'Auld Lang Syne' without adding a touch of Chinese twang. I wonder how many times they have dined out on the story.

I never saw or heard of them or the ship again.

17. *Sha He,* 1984

In 1984, the Republic of China State Shipping Company inaugurated a new service of top-class container vessels to compete in the far eastern to northern Europe trade route. The Chinese authorities were aware of, and concerned with, the fact that the masters and officers of these ships were unfamiliar with the procedures, rules and regulations, weather and the navigational and general operational difficulties which would have to be dealt with in the area of the English Channel and northern Europe. As a result, Trinity House was approached with a request to supply suitable and licensed pilots from the London Channel Pilots District to be on board those ships throughout this critical period.

At 0930 on 21 November, 1984, I climbed on board the *Sha He* in Gravesend Reach, to take the ship through the Thames Estuary to the Sunk Light Vessel and then on to Hamburg, Rotterdam and Antwerp. The weather was fair enough for the time of year – wind SW force 7 – and, on reaching the bridge, I knew that I had a very fine ship on my hands.

The captain and officers appeared to be affable and efficient enough, but their English was limited. However, helm and engine orders and general sea terms were well understood and, by the time we were passing Thames Haven and Canvey Island and entering the Sea Reach, I was on good terms with everyone.

I had the impression I was on board a highly sophisticated ship, that both the captain and the officers had total faith in my ability and expertise, that the navigation of the vessel was in my hands and nothing I did was to be questioned. This situation suited me well.

We cleared the Sunk Light Vessel at 1310 hrs. The weather forecast was not good but I was not too worried about that. It had been a routine passage up the Black Deep, and a clear run to the Inner and Outer Gabbards. From there, I set the course, allowing for wind and tide, for our next point of contact, which was the Texel Light Vessel, 88 miles away.

The bridge was well populated. I had, by this time, a good relationship with the captain and the officer of the watch, but the remaining six or seven men were a mystery to me. Everyone was dressed in the same way, except that the captain had black shoes and the others thick-soled sandal-type footwear. Their plain tunics were devoid of gold braid.

Food was mentioned and I immediately began to think of a typical Chinese meal. The steward arrived on the bridge, complete with white cloth, and my hopes were high. My first course, which was brought to me with a certain diffidence, was soup, and very excellent soup at that, full of vegetables and very nutritious, if a little glutinous. The bread, rice and fruit which followed were also acceptable, but the main course was not a success. It consisted of four fried eggs, cold and rubbery, and a type of spam which tasted like nothing on earth.

Eight pairs of eyes were watching my every move, so that I felt I had to eat and appear to enjoy what was on offer. Later, I found that the cook had never before catered for a European and that the galley had their own opinion as to what a European should eat. As long as the food was not tampered with – tomatoes, lettuce, fruit and so on – it was fine. The rice and soup were always very good, but when the cook decided to experiment with what he considered I should really enjoy, disaster followed.

Our passage across the North Sea was without incident. We

crossed the Deep Water route and, although traffic was heavy and the course had to be altered many times, I was not hard-pressed. I remained on the bridge throughout and was quite prepared for the long hours as I knew I would have to remain alert until we picked up the German pilot at the mouth of the Elbe.

The captain went below and I explored and investigated everything which was within my range, closely followed by three or four Chinamen who were always smiling and as helpful as they could be under the circumstances.

About 1700 hrs, the captain appeared and approached me in a very positive manner, holding two books and smiling as always. After some time, I found his intention was to discuss the operation of the very sophisticated radar and, in order to do so, he had armed himself with an English-Chinese dictionary as well as the operation manual for the radar.

It was a difficult situation. In the first place, our conversation was extremely limited and in the second place, my knowledge of the inner workings of the radar was somewhat sketchy, although the actual setting up, interpretation and operation of the radar was well within my ability to explain. I quickly found that we were not getting very far, but the captain seemed very pleased, and I passed a very pleasant hour with him, the two books and the radar. I probably learnt as much as he did, and it did have the effect of further cementing our relationship.

The weather was deteriorating and, by 2000 hrs, when we were abeam of the Texel Light Vessel, it was blowing a full gale from the south-west and the visibility had noticeably reduced. I altered course to pick up the first of the buoys marking the TX route along the Dutch coast and, by 2300 hrs, we were steering east once more on the TE inside route towards the Elbe.

During the run across the North Sea, food was again mentioned and, remembering my previous experience, I enquired as to the nature of the meal. The captain gestured with his hands and fingers in what looked like the action of breaking prawns, and my hopes were high. I was therefore dismayed when my immaculate steward brought me a bowl of what appeared to be washing-up water in which several boiled eggs were swilling around at various depths. The Captain had been illustrating the breaking of eggs rather than the preparation of prawns. I decided then to stick, if possible, to simple dishes and to experiment with great care.

The ship was behaving well, considering the weather, and we made good progress, passing the Borkum Riff Light Vessel at 0220 hrs, and pressing on to make the Elbe I Light Vessel at 0700 local time. I contacted the German pilot cutter on the VHF and passed my ETA (estimated time of arrival) at Elbe I, but I was informed that the cutter was off station due to the bad weather and had moved inside to No II buoy.

This presented no difficulty to me. I had taken ships inside the Elbe before and was familiar with the procedures. I had a limited area to turn the ship beam on to wind and sea to facilitate the pilot's boarding, but I was comfortable with the situation and, in due course, the German pilot appeared on the bridge and we proceeded up the river to the docks at Hamburg. Hamburg is one of the busiest ports in Europe, and ample evidence of this surrounded us on our passage to the berth.

After we secured alongside, the usual formalities of entering the ship into the port and completing the necessary documentation for customs, immigration and so on was carried out in the captain's cabin. By this time, I had been accepted as, if not a genuine member of the ship's company, at least a valued addition, and the captain felt I had the interest of the ship at heart. I found myself acting as interpreter and general assistant and, together with the ship's agent, who was an excellent fellow, we completed the necessary forms without delay.

I had been on Chinese ships many times before and always found that, apart from providing beer, the general hospitality was rather limited. However, I was amused to find that this had changed.

The German police and customs were plied with strong mint sweets placed in bowls as well as excellent Chinese brandy, which certainly helped the procedures and had a marked effect on the shore people.

The ship was due to sail at midnight after discharging a great many containers. By evening, I was very tired indeed, having had less than three hours' sleep since leaving London. I determined to turn in early in preparation for taking over from the German pilot at approximately 4am the next day.

My cabin, complete with en suite bathroom, was next to the captain's cabin and in the early evening, there was a knock on my door. It was himself. I invited him to come in and sit down and we talked, after a fashion, for a time, before he leapt to his feet and began to pace the deck. I could see that he was worried, but could not even guess the reason for his concern.

Eventually, he said, 'Mr Pilot, what about chips?'

My thoughts immediately flew to food and I had visions of the cook trying to produce his version of chip potatoes.

'Captain,' I said, 'I am very well pleased with the food. The steward and the cook are looking after me admirably and, if you don't mind, I would rather not have chips.'

'No, no, chips. Chips for the steward and the cook, chips. Chips, Mr Pilot.'

Our discussion went on for some considerable time until, to my astonishment, I found our conversation had nothing to do with food, but had everything to do with remuneration for the cook and the steward as a reward for their services to me. The Captain had been trying to say 'tips' rather than 'chips'. It had at no time crossed my mind to offer tips to communists on their own territory, but I was delighted to comply with the suggestion. I lost no time in doing so and, from then on, I could do no wrong. The ice was well and truly broken.

We sailed on schedule and I slept soundly throughout the departure, having left orders to be called half an hour before the German pilot disembarked.

I arrived on the bridge to find the wind screaming around us and an anxious pilot watching the heavy sea. He told me he would have to leave well inside the river or be over-carried to Rotterdam and needed my compliance. I was able to set his mind at ease. The wind was WSW force 8, and the pilot left the ship with some difficulty. In fact, I was surprised he didn't have to leave earlier.

We were abeam of Grosser Vogelsand and proceeded passing Elbe I Light Vessel at 0720, and set course for the inside route to Rotterdam.

The passage to the Maas river was anything but comfortable. The wind increased to force 9 and then 10, and our speed dropped from 17 to 11 knots. The visibility was bad, with heavy squalls of rain and the traffic was heavy. The seas were breaking on the bow, and spume and spray were everywhere. We eventually passed the Texel Light Vessel and altered course to 185 to make the approaches to the Maas.

When I called Maas approaches on channel 20 VHF to pass on my ETA and all information about the ship, I was told that all pilotage was suspended due to the weather. At 2330, we passed MN4 buoy and the outer anchorage. Lights were everywhere. Many ships were steaming slowly against the weather, unable to pick up pilots. Buoys were flashing and winking and the shore lights formed a blazing backdrop. A few ships had anchored, but some were dragging their anchors and were getting under way again. The situation was chaotic.

I contacted the authorities again, told them who I was and that I was preparing to come into the river and that I would require a Dutch pilot at the Hook of Holland. The port authorities acknowledged my message and replied that I should turn to starboard in the inner anchorage and steam for five miles and wait. Some time later, with the ship rolling heavily, we were told to come in and to listen out on channel 20 VHF for information and instructions.

By this time the captain was showing signs of nervousness, especially as the containers on deck were vulnerable to the movement of the ship. It was with a sense of relief that I ordered the large alteration of course necessary for the approach to the port.

I had foreseen the possibility of being unable to pick up the pilot and had worked out the tidal direction and strength along with other details I needed to control the ship; but I knew my reactions to the changing circumstances on the run in would be of paramount importance.

I had never before taken a ship in to the Hook of Holland, but, despite the weather difficulties, I was confident enough. There were lights all over the place but, as we dropped the entrance buoy and the other ships astern, I concentrated on the shore navigation lights as the adrenalin began to pump.

The entrance to the River Maas is not difficult. Two very powerful leading lights, which had to be kept in line, guided the ship through the narrow entrance and a further two red leading lights led the ship into the river. The lighted buoys before the entrance were useful, but my greatest problems were wind and tide. I knew that I would have to take great care of a very strong north-east setting tide and the wind was screaming out of the south-west.

I had passed the Maas Centre buoy on the starboard side, lined up the leading lights and was ready to begin my run in. The ship was making more speed than I would have liked, but I needed the speed to steer accurately. Very soon, I found the ship was being sent over to the north-east and I would have to counter this effect to stay in the channel. I altered 10° off course and then 25°. The ship was literally crabbing up the channel, and my Chinese friends were beginning to get agitated.

All this time, the shore authorities were feeding me information and advice, which I could have well done without, however commendable their intentions. It tended to distract me and filled the wheelhouse with noise. I did not dare switch off in case I missed something important, but I would have been better served had I been left to request advice.

Anyway, the breakwaters were rushing towards me at an alarming rate and I was well aware that the most dangerous part was still to come.

I could now see the two red lights on my port bow and I knew very well that I would have very little time to line them up and reduce the speed. I was also very much aware that I would have to make a large alteration of course to port and that I would lose the effect of wind and tide as the ship passed into the shelter of the river entrance.

The atmosphere on the bridge was not without tension. The captain seemed a little nervous. I had learned long ago that I should never show uncertainty, so I tried to appear supremely confident, especially as my every move was being watched.

Suddenly, we were very close to the northern breakwater. The time had come for fast action. The ship sheered to starboard, but I was expecting this to happen and I had already ordered the helm hard aport and the engines to stop. The ship surged on and I steadied her on 095 to bring the red leading lights in line and get the ship back in the channel.

This was not easy and, as we swept past the southern breakwater end, the helmsman was struggling with the steering. Our speed was reducing fast, but I was a little worried that the pilot boat may have difficulty catching up with us. We were soon abeam of the Maas pilot station and, to my relief, the pilot boat dropped alongside without effort and I was able to greet the Dutch pilot and hand the ship over to him with all systems completely under control. The time was 0400 hrs and my work was finished. We tied up at the container berth at 0600 hrs local time, but by that time I was fast asleep.

I slept soundly, aided by a large tot of Chinese brandy and when I surfaced it was lunchtime and I was very hungry. From past experience, I was not hopeful of an appetising meal, so I decided that, instead of dining in the isolated splendour of my cabin, I would go down to the dining saloon and take a chance on real Chinese food. This decision was not a success.

My entry into the saloon was greeted with stony silence and 20 pairs of eyes turned towards me. It was obvious that this was an extraordinary event. I persevered, however, and was rewarded with a meal which was infinitely worse than anything I had previously experienced.

We sailed at 1830 next day, bound for Antwerp, and the pilot handed the ship over to me at 2030, when we were just clear of the breakwaters. I immediately informed the Dutch authorities that I would be leaving the channel and steering for the Goeree Light Vessel, and we worked up to full speed, making for the Steinbank and Flushing Roads.

The passage was a pleasure. The wind had eased to a moderate breeze and visibility was excellent. It was a delight to be at sea. At 2245 hrs, the pilot was on board and we ran over the banks with plenty of water under the ship. We changed pilots at Flushing and tied up in Antwerp at 0800 hrs on 26 November.

My tour of duty was over and I left the ship in Antwerp to return to London. The ship was to remain in Antwerp for several days and another pilot would take her on down the channel. I had made a good friend in the captain, and the steward was nearly in tears when I left. It had been an interesting few days, with plenty of excitement, and I felt quite sad to be leaving.

A few months later, the *Sha He* docked in Tilbury and, to my surprise and gratification, the captain demanded my services to take the ship to the Continent. Unfortunately, our rules dictated that the first pilot on the roster with the appropriate licences should be offered the job and, as I was on leave at the time, the captain could not be accommodated. However, it was a great compliment, especially when the pilot who had been assigned to the ship telephoned and told me that the captain had been loud in his praise of 'my special pilot from Scotland.'

18. *Arroyofrio Uno*, 1987

On 14 January 1987, extreme, very severe weather was evident in the South East of England. Heavy snow had fallen, blizzard conditions prevailed and the wind was rising steadily. All transport – cars, trains, buses – were unable to move, and it was intensely cold.

The pilot station at Gravesend was faced with a desperate situation. Many pilots who lived outside walking distance from the pier were confined to their homes. The conditions for boarding and landing to and from ships were exceedingly hazardous. The duty pilot, in conjunction with the Port Authorities, decided to close the pilot station until the weather improved. This decision was serious indeed. To my knowledge, Gravesend Pilot Station had never closed and we prided ourselves that a pilot would be available for any ship requiring his services.

The river traffic was also naturally very sparse. Many sailings had been cancelled and, when the pilot station was officially closed, only one ship was preparing to sail. She was a small Spanish RO-RO ferry named *Arroyofrio Uno* and she was coming from Deptford and bound for Zeebrugge.

On the afternoon of 14 January, I received a telephone call from the duty pilot, who said, 'Jim, you are not yet No 1 on turn, but you are the first pilot on the roster who has any hope of getting in here. We are closing the station temporarily but we have one ship requiring a pilot, and it's only fair to put you in the picture.'

My reaction was immediate. I told him I would get kitted out, sea boots and heavy weather gear, and I'd be there within the hour.

I set off walking to the pilot station, keeping to the middle of the road, stepping with great care, and met nothing on the way. When I arrived, I found a very grave duty pilot who was not at all happy with the situation.

The *Arroyofrio Uno* was still on the way downriver and in VHF contact. The river pilot was obviously anxious to disembark at Gravesend and the captain was determined to go to sea if he could get a pilot.

I spoke with the pilot boat skipper and found he was willing, if reluctant, to attempt a boarding. I thought that the *Arroyofrio Uno* captain must be mad to go to sea, but suspected he was under pressure from his owners to keep the ferry service going at all costs.

Anyway, with the help of the crew of the pilot boat, and after a few hair-raising moments of trying to get alongside the ship, I was finally able to climb the pilot ladder and gain the deck.

On the bridge was a very nervous Spanish captain and helmsman. The river pilot had positioned the ship as best he could to give some shelter to the side on which the pilot ladder was positioned, and I was able to do the same for him, and he disembarked safely. Then I rang the engines to full ahead.

Gravesend Reach and the Lower Hope Reach were both successfully navigated and, as we cleared Shell Haven and entered the Sea Reach, the full power of the wind was very much in evidence. Visibility varied between nil in blizzards of snow, to good between the squalls. The sea was very rough and the entire scene was more like the Arctic than South East England.

The ship was responding well, but it was bitterly cold – due apparently, apart from the weather itself, to the freezing-up of the heating systems on board. The weather had affected many of the onboard facilities and conditions were far from pleasant.

We passed Southend and the Red Sand and Shivering Sand Towers, and gained the open sea through the Princes Channel. It was not the easiest job of my career, but all went well.

Services had long since been suspended by the sea pilot stations, so I had no option but to proceed with the ship. My work was finished at the Tongue Light Vessel where I handed over to the captain and went below. I enjoyed a reasonable meal in the saloon and was shown to my cabin, where I found a double bed instead of the usual bunk.

I was not too happy with the sleeping arrangements, given the hazardous weather conditions. A double bed in the centre of the cabin with no means of preventing me being flung out on one side or the other was not my idea of comfortable facilities. I would have preferred a traditional bunk with a bunk board on one side and the ship's bulkhead on the other so that I could wedge myself in. But I had no option and resigned myself to my fate.

I spent a very uncomfortable night, being tossed about in bed and finding myself on the deck on more than one occasion. In the early morning, I gave up and, having dressed, made my way to the bridge to see what was going on. By my reckoning, we should have been within striking distance of Zeebrugge.

My professional services had finished at the North East Spit buoy and, officially, I was a passenger, but it was in my interest to offer what help I could and I was happy to do so.

The atmosphere on the bridge was tense. The weather had improved only a little and the North Sea was in a vicious mood. There was no sight of land, and all there was to be seen were the battalions of waves marching on the ship and causing her to pitch and roll without respite. I entered the chartroom with the captain and found that we were in the east-going traffic lane and were well on our way towards Rotterdam.

The Captain informed me that he had been forced to put the wind and sea on the bow for the safety of the vessel and that we had been slowly steaming northwards for most of the night. We were, in fact, on a course that was taking us ever further from our destination, and so the captain and I went into consultation.

The weather, although still fearsome, had improved slightly and, after calculating the tide and watching the sea for a while, we decided on a positive plan. I suggested we should all have breakfast first and then go into action. We would warn the ship's company and, when the time was right and the ship was in a comparatively calm area of water, which came along from time to time, we would put the helm hard over and, using the engines, ease the ship round until the wind and sea were on the quarter, and then run directly for Zeebrugge across the sandbanks.

The tides were right and, with our comparatively shallow draught — 17 feet — we could cross the Thornton Bank, picking up the Thornton buoy, cross the Westpit and the Schooneveld, sighting the Akkaert buoy to port, and we should be in Zeebrugge in a couple of hours. And that was exactly what we did.

It was exciting and exhilarating sailing and, apart from some broken crockery in the galley, we accomplished our aim and triumphantly entered the harbour at Zeebrugge.

The port was in desolation. Snow, ice and slush were the predominant features and everybody was miserable. The next few hours were spent discharging cargo and defrosting and freeing the essential equipment on board the ship.

I remained on board as the ship was due to sail that night back to London and, with the wind easing to moderate, we put Zeebrugge behind us and, after an uneventful crossing, picked up a pilot at the North East Spit buoy.

I was a little disappointed because, had we been unable to pick up a pilot, I would have been free to take the ship up the river and claim two acts of pilotage instead of one. However, I was very glad to be back in Gravesend and to get home to a hot bath and a long sleep.

19. *San Salvadore, 1987*

On 18 October 1987, I boarded the Spanish cargo ship *San Salvadore* off Gravesend to take her to sea. A few hours later, the south of England was to be savaged by a phenomenal storm, the most savage sequence of weather in living memory.

At 1600 hrs on that day, when I climbed to the bridge of the ship, the weather forecast for our area was bad, but not unusual – wind SW 8–9. Folkestone Pilot Station, where I would normally disembark, had already suspended service but Margate Pilot Station was operating a restricted service, so I was not unduly concerned.

The *San Salvadore* was well found and efficiently run. The captain and officers were pleasant and very hospitable. We set off down Gravesend Reach and turned into the Lower Hope, and it was noticeable that the force of the wind was increasing. By the time we were off Southend, the spume was being blown from the tops of the waves and all the indications were of a dirty night ahead. The flood tide had been running for three hours by the time we were passing the Red Sand Towers, so the Princes Channel was the obvious choice to clear the estuary.

I was in touch with the pilot station ashore and had arranged to bring the ship close in to the shore off Margate pier and, with the wind at SSW, I foresaw little difficulty in disembarking.

The passage through the Princes Channel to the NE Spit buoy was accomplished in increasingly heavy weather, which necessitated careful handling of the ship, especially as the inward-bound traffic was considerable.

I turned the ship into Margate Roads and reduced speed as we approached the pilot boat. The sea was appreciably less as we gained the shelter of the land, and the boat had no difficulty coming alongside on the port side where the pilot ladder had been rigged.

I gave the captain the course to steer after I had left the ship and instructed him in detail how he should proceed to the North Goodwin Light Vessel, then into the south-going shipping lane. As the pilot boat pulled away, I waved farewell and watched as the ship set her course to Goodwin and on to Seville. I never saw that ship again, but often wondered how she fared in the hours to come.

I landed on Margate Pier, battling against an ever-increasing gale, but still did not anticipate any extraordinary phenomenon. There had been no indication from the weather forecasts that anything other than seasonal weather was imminent and I was looking forward to a quick dash home on the last train and a quiet night in my own bed.

The train was very late when it arrived at Gravesend. The delays were caused by obstructions on the line from the ever-increasing wind force. My wife and family were in Scotland visiting relatives and I was aware that no warm welcome would be waiting for me. However, I was very tired and lost no time in getting organised for bed and was asleep almost before my head touched the pillow.

I awoke with a start at about five o'clock and was immediately aware that something very strange was going on. My first impression was of being surrounded by noise – bangs and crashes and, above all, a continuous high-pitched whine, almost a shriek, which was very disturbing.

The house seemed to be shaking but, on reflection, I think my imagination was at work. I leaned over to switch on the bed light, but found that there was no welcome light from that source. It was pitch dark. I tried the light switch at the door, with similar success. I pulled on a dressing gown and went over to the window. Normally, I would have seen many lights because the view from the window took in a large area of the town down to the river and most of the Lower Hope Reach from Shornmead to Thames Haven.

It was a perfectly clear morning, but lights were few and far between, so I concluded the electricity had been cut off. I felt my way down to the kitchen and opened the back door. I stepped outside and was almost knocked down by the hurricane force of the gale.

I staggered across the patio and found myself on the lawn. A dangerous and frightening situation had developed. The wind was literally devastating the area. Our house stood on a hill and was exposed to the elements and, from where I stood, I was aware that slates, especially ridge tiles, were being lifted from the rooftops and were landing all around. Bushes and shrubs had been uprooted and half the wooden garden fencing was scattered around the lawn. I very quickly retreated indoors and decided to wait for daylight before doing anything else.

As time went on, I realised that the shriek of the wind was lessening. Later still, after the break of dawn, it became obvious that the worst was over, that the wind was taking off and, hopefully, some sort of normality would return.

Nearly all the houses in the neighbourhood had lost tiles and some had gaping holes in their roofs as well as other structural damage. Many bushes and even trees had been uprooted, but the electricity had been restored and I was able to make breakfast. It had been an exciting, if unpleasant, night.

I drove down to the pilot station later that day and it was as though Gravesend had been under bombardment. Branches of trees, tiles, bricks and all sorts of curious objects were strewn about, and I handled the car with great care. The pilot station was a hive of activity. All the ships anchored at Southend had dragged their anchors and were ashore, and were in immediate need of a pilot. The riverbanks were littered with small craft that had been driven on to the mud.

Gradually, the situation was brought under control, but it had been a difficult and dangerous night for all concerned with the river and its shipping. A large passenger liner, the *Stefan Batory* had had a very hard time indeed. Attempting to come upriver from the sea to berth on Tilbury jetty at the height of the storm, she had been informed that berthing at Tilbury was impossible. The ship had therefore attempted to turn round and gain sea room. During the resultant manoeuvres, she lost both anchors and the pilot and captain gained a few grey hairs. The *Stefan Batory* was successfully berthed later in the day.

The south of England suffered extreme devastation that night and the recovery process was difficult and protracted. The weather must have been the most severe of the century and I, for one, was thankful to have come through it comparatively unscathed.

20. *Aztec Lady* and the Tall Ships Race, 1990

The *Aztec Lady*

The *Aztec Lady* is a two-masted Bermuda rigged ketch, a very fine seaworthy yacht built for a millionaire who unfortunately died before he could enjoy her qualities. She was fitted out with considerable luxury for cruising, but not for racing, and had a powerful engine and all modern navigational aids. A friend of mine, John Heath, bought her. He had a bargain!

John Heath and I were old shipmates. We had trained together on HMS *Conway* then went our separate ways. I finished up as a Trinity House Channel Pilot on the London river, but John, after obtaining his Masters ticket, finished up as a very successful businessman and a Royal Naval Reserves commander.

The *Aztec Lady* was no stranger to the Tall Ships Race and, in July 1990, when John asked me to join him, I agreed immediately.

The race was to begin at Plymouth and the first leg was to be Plymouth to La Coruña in northern Spain. Next was a cruise rather than a race to Bordeaux in France. The second leg was the race from Bordeaux to Dunkirk. *Aztec Lady* was dropping out of the race at Bordeaux because of previous commitments.

Aztec Lady was to be crewed by a couple of friends, and girls from John's local schools. She would be officered by John, myself and another experienced yachtsman and engineer, and the cook was to be John's wife, Joan – a motley crowd indeed!

A great deal of work had to be done before the start of the race. Stores had to be loaded, her engine and every item of ship's gear had to be checked and the new crew had to be given basic training. The formalities of entering the ship and the final briefing, during which John fell asleep, had to be completed.

At last, the great day of sailing arrived, and we on board *Aztec Lady* were ready for anything. Ship after ship slipped their moorings and set out to take part in the Parade of Sail. The big square-rigged sailing ships from Russia, South America, Italy and so on were an impressive sight and fortunately the weather was kind.

The plan was that the ships, in order, would sail past HMS *Penelope*, a Royal Navy frigate, which was to be the communications and guard ship for the fleet and would accompany us on our voyage. She had a guest on board, Princess Anne, who would start the race but not come to sea with us. The ships would then proceed to the official vessel on the starting line and out to sea.

John Heath, who sported an impressive set of whiskers, had his crew lining the decks to cheer the frigate, while he leant out of the wheelhouse window and encouraged them. We were all highly amused to learn later that Princess Anne had expressed interest in the strange sight of *Aztec Lady* and her piratical crew, and was told she was one of the only two private vessels taking part in the race and was owned and skippered by an 'old Derbyshire goat' and his crew of kids.

After the ceremony, John decided that, instead of sailing immediately, we would anchor in a quiet spot and have a meal, which we did, causing consternation among the officials who could not understand what had become of us. The weather had turned misty, restricting visibility, and repeated calls for us on the VHF radios produced no reply, simply because our set was switched off and we were all tucking in to lunch.

Finally, John came to the conclusion that we had better make a move. We were the last ship to pass the starting point and our reputation as a ship of character was established.

The weather was working up to half a gale from the south west and *Aztec Lady* began to feel the pressure. Most of the crew were seasick and pretty miserable, but the work of the ship had to carry on, and we tacked our way across the channel to the French coast.

The fleet were, of course, prohibited from using their engines and, with the wind in entirely the wrong direction, progress towards our destination was difficult. The big square riggers were driven east up the channel. Each of the armed services had entered very fast ocean-going yachts and they seemed to be the only contestants well on their way past Ushant and making excellent progress.

Aztec Lady laboured on, tack for tack, and eventually, at the western end of the English Channel, John decided to go inside Ushant, which was an exciting way into the Bay of Biscay. But it was not an easy undertaking. There was a very strong, wicked tide, and rocks littered our course. We marked time until the tide suited us, then plunged south at high speed – an exhilarating experience.

The Bay of Biscay was in a comparatively benevolent mood and our crew, if not totally cured of seasickness, were much more comfortable. Joan Heath and her galley staff were producing excellent meals and, although we were not breaking any speed records, progress was being made in the right direction. We saw nothing of the other ships, but we did have two idyllic days' sailing, with dolphins playing around the bows and the wind backing into the WNW. The ship's company had settled down and, apart from one girl who never really adjusted, we were pulling together and enjoying the trip enormously.

I was amazed at the ability of these young people, who found themselves in a totally strange environment, and how they adjusted to being called upon to cope with lack of sleep, hard work and strange duties, especially when they were also subject to strict discipline and considerable discomfort.

We had, of course, seen no land until we raised the coast of Spain and the port of La Coruña was within our sights. We knew that we had not made a fast passage, and this was emphasised when we entered the harbour and found the port saturated with sailing ships, including the big square riggers. It was a magnificent sight, and we were greeted with ribald, if good-natured, shouts about our late arrival as we berthed stern on to the jetty in the inner harbour.

La Coruña was in a carnival mood. The area around the harbour was open to the public and packed with sightseers strolling around looking at the variety of ships and thoroughly enjoying themselves. We spent five days at La Coruña and the weather was perfect. Parties and jollification were the order of the day and the city fathers and officials did everything to make our stay memorable.

La Coruña

There were excursions to various locations of historical interest and of local interest, but the main event of our visit was a parade of ships' crews and local musical bands, culminating at the main public park, where a great buffet of food and drink was provided and the trophies and prizes awarded to the winning ships in each class. The Mayor and high officials were to carry out the presentations and everyone had a wonderful time. Our valiant crew cheered the big square riggers and others when they received their prizes, cups and trophies and, being aware that our racing qualities left a lot to be desired, we were content to be supporters.

After the main ceremony was over, the Mayor announced that he had the honour and pleasure to present a special award and that it was to be given to the sailing ship which had necessitated the cook to produce the maximum number of meals *pro rata* on the passage between Plymouth and La Coruña. It came as rather a shock when *Aztec Lady* was called, but Joan Heath and two of our

155

girls rose valiantly to the occasion, mounted the VIP platform and, to the accumulated cheers of the ships' companies, were happy recipients of a splendid, suitably inscribed butcher's cleaver.

Included in the many invitations to *Aztec Lady* and her crew was one addressed to the captain which requested him to take breakfast with the captains and officers of the two Swedish sailing ships, the *Gladen* and the *Falken*, both berthed not far from us. John Heath delegated me to represent the ship and it was not until later that he informed me that he had already attended a similar breakfast on a previous Tall Ships Race and it had taken him two days to recover.

In all innocence, I walked across to the ships. I had dressed suitably for the occasion and was looking forward to a couple of hours of pleasant conversation and good Swedish breakfast fare. I soon realised that Swedish hospitality went far beyond my expectations.

The ships were moored alongside each other. Their quarterdecks had been cleared and decorated with flags and bunting, and tables laden with food were set up on both ships. The two Swedish captains greeted me at the gangway and I was ushered into the hands of the Swedish Naval Cadets, male and female, who were acting as hosts and stewards. A glass of schnapps was thrust into my hand as an appetiser and, from then on, I was treated like a king.

Both quarterdecks were crowded with guests and I began to realise that breakfast on board these magnificent vessels was an experience of a lifetime, and probably never to be repeated. The food was of the highest quality and nobody seemed to be allowed to have an empty glass. The two captains were everywhere, attending to the visitors, and it was they who announced that we were now expected to sing.

The Swedish cadets produced screens upon which were written the words of the songs which were to be the accompaniment to our breakfast. Most of them were good-going drinking songs, easy to sing, and all included toasts and much quaffing of schnapps. It was enormous fun and, of course, everyone entered into the spirit of the occasion.

I left the ships at lunchtime and made my way back to *Aztec Lady* with a rather more unsteady gait than when I had set out. The crew were waiting for me, John Heath having briefed them on what to expect. I was escorted to my bunk, and was later told that the Swedish breakfast ended at 1700 hrs, but that the party continued until much later.

Another incident which caused great hilarity among the crew was the spectacle recovery. We had been invited to spend the day with an escorted party to sightsee the historical and interesting parts of La Coruña and surrounding areas, and most of us accepted. The engineer, however, decided to remain on board and to look after the ship.

To understand what followed, I shall have to explain our gangway arrangements and our rather precarious method of getting ashore. The ship was stern on to the jetty, with two anchors out forward, and we had rigged up an ordinary domestic ladder leading from the stern to the quay. The set up was anything but ideal. There were no handrails and to transfer from ship to shore, you more or less scrambled on all fours up the ladder. Fortunately, we were all fit and agile and had little difficulty, but it did nothing for our dignity.

I had placed my spectacles in the breast pocket of my shirt and alas, on my way up the ladder, my glasses fell out of my pocket and dropped into the dock. I was very annoyed, but I did have a spare pair, so it was not a complete disaster. The engineer standing on deck had witnessed the entire incident and was very sympathetic, and the girls comforted me as best they could.

We set out on a splendid day's outing and did not return until late in the afternoon. My amazement can be imagined when the engineer ushered me into the saloon, and there were my spectacles on the table, sparkling clean, unmarked and intact. The engineer was bursting with pride and I could not wait to hear the story of their recovery.

Apparently, the engineer, bless him, had taken a bearing on the quay wall and knew exactly the spot where the glasses had fallen into the water. With great presence of mind, he had called up the Royal Navy frigate HMS *Penelope* and poured out the whole story, pleading for help, offering them a free exercise for their skin-divers, and stating that his poor navigator would not be able to see the charts unless a search and rescue operation was mounted at once.

HMS *Penelope* responded with typical naval efficiency. Within a very short time, a team of divers, with their back-up crew and equipment arrived at *Aztec Lady* and set to work. They caused quite a stir and attracted a large crowd on the waterfront.

The dock water was anything but clear but, by following the engineer's bearings and directions, the divers recovered my specs without any fuss and raised a cheer from the onlookers. The Navy cleaned the specs thoroughly and, without any ceremony, presented them to the engineer before vanishing in their inflatable boat back to *Penelope*. Altogether, a very impressive example of co-operation and goodwill and, to this day, I think of the incident with gratitude.

The citizens and officials of La Coruña extended an almost unbelievable level of hospitality and friendship to the crews of all the sailing ships and we all appreciated their generosity. Among the many events they organised was a visit to the cathedral city of Santiago de Compostela and those of us who accepted the invitation were rewarded with a splendid day out and returned saturated in local culture and history.

Our five days in La Coruña were well spent, but inevitably the day of sailing arrived. The ships put to sea as a fleet, not to race, but to make our way to Bordeaux as we pleased and set our own pace, as long as we arrived in France on the designated date. This was to be a leisurely cruise and some of the ships had exchanged individual members of the crews, which allowed different nationalities to sail with each other and hopefully cement friendships.

The experiment was of limited success, and some of the British crews who had exchanged with foreign sailors were not overly impressed, especially on board the big Russian ships. Our crew turned the offer of exchange down flat, feeling they obviously knew where their best interests were centred.

The *Aztec Lady* at sea

Aztec Lady set her course along the north coast of Spain, and next day she found herself at anchor, close inshore in the bay of a small Spanish town. We spent the day in perfect weather, swimming and basking in the sun and generally enjoying a short period of rest and recuperation.

We weighed anchor late in the afternoon and headed north for Bordeaux. We occasionally sighted some of the other ships but it was not until we arrived at the approaches to the Gironde river that we began to converge and move on up to the port. The berths at Bordeaux were all along the riverside and the sailing ships made a grand spectacle lined up stem to stern. Decorations and flags were everywhere and the fair city of Bordeaux had done everything to make us welcome and make our visit memorable.

French hospitality rivaled the Spanish in its open-handed treatment, and we were again inundated with invitations to parties and excursions. Enjoyment was the order of the day.

One of the many diversions was an organised trip to a vineyard outside the city. We travelled by coach, and the vineyard had laid on a lavish buffet lunch in the grounds of the chateau which, of course, included as much wine as we could have wished.

At the end of the day, the owner presented each of us with a bottle of red wine, with instructions that the bottle had to remain unopened for five years. Our crew immediately made a pledge that, in five years' time, we would have a reunion and open all the bottles. The fact that the reunion never did take place was regrettable, but circumstances made it impossible – the best laid schemes, and all that.

The culmination of a very happy stay in France was spectacular. On the night before the fleet sailed, when everyone was partying on different ships and the quays were alive with sightseers, the night exploded into colour as a magnificent firework display was put into action.

It was a fitting end for the *Aztec Lady* trip. Joan and John Heath were about to drop out of the Tall Ships Race, welcome members of their family as new crew members and have a leisurely cruise to the Channel Islands. The girl crew and I flew home next day.

I was to be linked once again with the *Aztec Lady* at a later date.

But that is another story.

21. Afterword *by Alan Riach*

There are always other stories, waiting to be told. The experiences people have in all walks of life are not equally represented in literary work. Some professions seem to lend themselves to certain forms or genres and tales of the sea and ships have a particular and prominent place in the history of world literature, from Homer to Melville and Conrad.

My father wrote his stories, he said, not for publication but simply so that his grandchildren might have some idea of his life, things that happened to him, the world he came from, the priorities he assumed and lived by. The first four chapters of this book are based on transcriptions of recorded conversations I had with him between August 1998 and January 1999; chapters 5 to 20 he wrote himself in the 1980s and 1990s. The whole book was lightly edited by Carl MacDougall, to whom my father and I are very grateful.

A handful of phrases he would use from time to time stay in my mind. When he boarded a ship at night and had to wait for the tide, he took advantage of the opportunity of a few hours' sleep before coming on deck to do the work of piloting. He would turn in with the line, 'Wake me when she swings...' The ship would swing round in the turn of the tide and that would be the signal for his work to begin.

At home in Gravesend, in Kent, when I was a boy, there was never any semblance of a 9-to-5 working day. My father would be called to work at any time of the day or night, according to the duty roster at the Pilot Station. He would normally get an hour's notice by phone and have to get himself to the Station in full

James Riach going to work

uniform ready for work within the hour. He would come home at any time, usually tired and ready for sleep, and we had to respect the need he had for rest at such times. Everything must be kept hushed in the house while he was sleeping and we would never wake him. He would waken himself when he had had sufficient rest. As a teenager, I remember when he would come home with a bottle of whisky, a gift from the captain from the bond locker of the ship he had been piloting, and a dram would be taken before he turned in. He would occasionally quip, on his way upstairs, 'Wake me when she swings...'

Perhaps partly as a result of the telephone always being a machine that delivered the order to work, when he was on leave and after he retired, he was normally negligent of its ringing. My mother, sister and I would almost always answer the phone before he did. He seemed to be disdainful of any claim it might have made upon him, when he wasn't on call. Close focus on matters of immediate concern was characteristic, excluding extraneous things altogether. I have never met anyone who could be so oblivious to what he deemed unimportant things and yet so comprehensively aware of everything significant around him. Once, he was watching a television programme about shipping and ships on the small TV in the kitchen. I had arrived home in the evening and put two slices of bread under the grill to toast, when the phone rang. I went to the hall, closing the kitchen door to block the sound of the TV, assuming he would keep an eye on the cooker. I had talked to him and walked past him a few times while I'd been in the kitchen and it didn't seem an unreasonable assumption. The phone call kept me busy until I saw the smoke coming out from beneath the kitchen door. I hung up, went through and found the toast black and blazing, my father totally oblivious to the cumulo-nimbus charcoal clouds billowing out from under the grill. He was surprised when I pointed out what had happened.

My mother figures in this narrative mainly in the light of my father's home life and the background to his professional work but it would be wrong to leave her own story unacknowledged, and he would not wish to do so. He refers to his marriage with great affection and respect for my mother. It was not a subject he ever talked much about and never with romantic excess. A few times, however, I remember him saying that upon entering married life he had become gratefully aware of the shelter such a condition afforded him. He had, he said, arrived 'In the lee of Bum Island...'

He repeated this once too often, in my mother's hearing, and I remember the sensitivity with which he realised he had said something a little too vulgar for my mother's sensibilities. He didn't repeat it.

She was born in Salsburgh, Lanarkshire, on 23 August 1928 and moved with her parents to Calderbank in February 1932. She was educated at Calderbank Primary School from 1933 to 1940 and then Airdrie Academy from 1940 to 1946, gaining eight Highers and going on to Glasgow University, from 1946 to 1950, graduating with a B.Sc.(Hons). She then spent a year at Jordanhill Teachers' Training College before joinging the National Coal Board as a Physicist and being appointed as a Mathematics Teacher at Kildonan Secondary School in Coatbridge and then at Airdrie Academy.

She was the second child of a large family of five sons and two daughters – John, Janette, David, MacArthur, Alex, Glen and Margaret. Janette was a member and regular attender of Calderbank Parish Church, a member of the Girls Guildry and later a Sunday School Teacher. She met my father in Glasgow and they were married in April 1956, in Glasgow University Chapel, by the Minister of Calderbank Parish Church at that time, the Reverend Gilbert Mowat. Janette continued to live with her parents in the building next door to the church, at 20 Main Street, Calderbank, while my father was at sea. I was born on 1 August 1957 in Airdrie House and we lived in Calderbank before my parents moved to Gravesend in Kent in 1961. I stayed on with my grandparents a while longer before moving south. My sister was born in Gravesend in 1962 and my parents lived happily there for thirty-six years, though at every opportunity, they headed back north to Scotland almost every time my father had some leave. They – or, we, usually – stayed most often with my grandparents and then my grandmother, first in Calderbank and later in Strathaven, in Lanarkshire.

Jimmy and Janette at *Strongarbh*, James Riach's childhood home in Tobermory, Mull

My parents returned to Scotland permanently in 1997. They bought a house in Hollybush, Ayrshire, and eventually moved down to Alloway, three corners away from Robert Burns's cottage, and four away from me, my wife Rae and our children James and David.

My mother was an excellent cook and a keen gardener, especially in Kent with its sunny climate. She made a very good Japanese Rice Wine – Saki – but, trying to utilise all resources, on one occasion discovered that after distillation the rice wasn't much use for rice pudding so she threw it out onto the back lawn where the birds had an alcoholic feast, ending with them drunkenly flapping against the kitchen windows!

She was a preserver of the past, keeping her own and her children's school maths books and English jotters. She loved her own parents dearly with a particular affection for her mother, and during the last few weeks of her mother's life, she spent many hours with her, gazing into the past together, eating ice cream and remembering all the good things of their time together. But all the Cunningham family were very close.

My mother died of complications brought on after years of vascular dementia on 27 November 2012. She had forgotten so much, yet nevertheless, from time to time, when I visited, I would see her hold out her hand and gesture affectionately to my father and call him over and hold his hand and tell him she loved him and he would say he loved her too and in those times when her mind was clear the love and affection they knew together over so many years was full and present and filled their house.

When she died, and my sister and I were working through the details of paying for the necessary expenses, my father, watching over us approvingly, sighed and said quietly, 'Spend whatever is required, because I have just lost someone who meant more to me than all the money in the world...'

It is only right to let my mother's presence be felt presiding beside him, for in all of the years his stories cover since he met her, there was never a trace of rivalry or rancour between them. One moment I recall was after a grand dinner, the discussion in full flow, after I'd returned to the parental home from a long sojourn elsewhere, and all was storytelling, convivial questions and long elaborations of accounts of our respective experiences, and at one point late in the latter part of the evening, my father held up his whisky in its cut crystal container and commented to my mother,

'You know, Janette, this is a *good* whisky glass!' His register of pleasure was emphatic and undeniable. My mother shook her head and gently, firmly, said, 'Jimmy, it's a half-pint tumbler!'

My father's parting lines, at the end of such evenings, or in the many long-distance telephone conversations or at the innumerable leavetakings at airports or railway stations, were equally memorable. 'Keep your powder dry,' he would advise. Or, 'Keep your topknot on.'

The steadiness of calculation, the human insight and knowledge of the ways of the natural world at sea, the sense of rightness, fairness, responsibility and fun, comes through in his stories, not only in what they describe but in the style of their writing. The formalities of understatement are frequently observed. They are works of craftsmanship and fidelity. When I was a wee boy, my father gave me *Huckleberry Finn* at Christmas 1967, when I was ten, and *Moby-Dick* at Christmas in 1972, when I was fifteen. I still remember the pleasure of going along the Mississippi with Huck and Jim, that first time, and the feeling when the breath left my body, literally, when Ahab goes over the side. The writing helps.

So do all the arts. My father was always interested in painting, and his own water colours of sailing ships were an aspect of his life that acknowledge his practical engagement with trying to see things through other forms of imagining and presenting them.

In later years, I wrote a number of poems that either drew directly on my father's stories or reflected on them obliquely. One, the second of the three 'Passages from India', is entirely given in his own words. Each connects with moments or locations in the stories in this book. Here they are.

My father talked to me once about an early voyage when he sailed past Lisbon, and the ship didn't go in, but the allure of the place and the scents drifting out from the land to shipboard were tantalising. In Australia, uncertain in my own memory whether he had referred to Lisbon or Adelaide, I wrote this.

Another Scent of Distance

'The perfume drifted out upon the waters,'
my father (he was first mate on a Clan Line
cargo vessel manned by lascars bound
for Sydney) told me, &
'We couldn't go in, though we wanted to
go in, it looked so
sunny & the trees were
dancing silver on the skyline &
the green leaves & the green grass shimmering.'

Adelaide, or Lisbon?
For now I can't remember
which route it was he would have taken, & know
I make a fiction of it all, like this.

Standing among the trees at Bedford Park,
looking past the blue gums down the hollow of the ranges,
to where the city stretches, over the wide estuarial plain,
low below a hovering cloud of blue polluted air,
I can breathe the scent coming out of the places around me,
& make of it this fiction I believe: I can
watch it go in colours from these Flinders ranges
drifting down the howe towards Adelaide's broad streets,
its stone-built buildings, pleasant & relaxing in the sun,
& out upon the waters where I do not know
my father may have passed, so long ago.

Growing up in Stornoway, my father accompanied his parents to church regularly every Sunday. The Church of Scotland service was normally preceded, he told me, by a service for the fishing men and their families, who would sing the hymns collected in the Moody and Sankey Hymn Book. Ira Sankey and Dwight Moody were American evangelists of the late nineteenth century, who met in 1871 and travelled through the United States and the United Kingdom, arriving in Scotland in 1875, and attracting

The Wind Ship, by James Riach

The yachts *Britannia* and *Valkyrie*, by James Riach

Bound Away, full rigged ship *MacQuarrie*, by James Riach

huge crowds. Moody preached and Sankey led the singing. Their rousing enthusiasm was taken up in later years by such eager evangelists as Billy Graham. I once found a battered second-hand copy of a Moody and Sankey Hymn Book and bought it for my father. He seemed to recognise it, as if it might have been one of the very copies he had seen used in the church in Stornoway. A long way from pious submissiveness to orthodox authority, I imagine the fierce feeling of those fishermen expressed in song. This poem was written thinking about that.

Sankey Hymns

What did they know of reverence,
the sailors' congregation? In Stornoway
they gathered, an hour before
the regular service
began, and filled the wooden pews and raised
their voices in mass praise. Resolve
of hardihood was theirs. Let reverence be left
to those more pious worshippers
who occupied the space left vacant when these men
went down to the sea to their ships, once again.
It's not a skill that's needed, catching fish.
Accuracy, yes. And Strength. A physical fact so prevalent
the eyes and voices of these men reveal it
every moment. Thinking was too slow for them,
reverence unhelpful.
    Let them stand in Stornoway,
and sing for God to help them live and work.

When he was a pilot and we were living in Gravesend, when I was in my teens and early twenties, sometimes we would be notified by phone at home that my father was returning from a voyage and we could drive down to one of the ports on the Kent coast and pick him up ourselves. This would save him the trip home by train and would give us an opportunity to drive through Kent sometimes at night, sometimes on sunny afternoons, as in this poem. Occasionally other pilots, friends and colleagues of my father, would be coming back in at roughly the same time and we would pick up all of them and fill the car. And sometimes, if the timing was right, we would break the journey by calling in at a pub on the high plains of Kent called 'The Long Reach'. The name of the pub stayed in my mind as a metaphor worth keeping.

The Long Reach

On the green plains of Kent, on the Canterbury Road,
under the highest arches of sky and the big silver banners of cloud,
where the long low marshes of the river's south bank are north-west,
bordering parishes, churches, graveyards, where Pip's young siblings'
    lozenges
lie in rustling leaves at Cooling, and every headstone's shelter hides a
    Magwitch,
where, to the north, Whitstable skims on its mudflat out into the
    estuary's birdsong and
where, south-east, the sheer vanilla stone of the Cathedral's pilgrims
    navigate
the Canterbury crowds, and where, all round the Cinque Ports,
    Margate, Ramsgate,
Deal and Dover, Folkestone, the pilot boats go out and come back in,
and the pilots look back on the grazing sheep on Romney Marsh,
near the great almost luminous ball of the power station at Dungeness,
and two oak trees on the curvature of skyline together make a rearing
    horse,
while here, on the green curved plains of Kent, we park the car
beside the pilgrims' road, by a long, low pub, and stand beside it,
looking all around, from Rochester downriver to the crumbling cliffs
in the south, then back up the river to London, and turn and look up
    once again,
reading the name on the tiled roof there, and then lock the car, and go
    in.

When the pilotage service was subjected to restructuring at the time of the Conservative government of Margaret Thatcher, my father felt the destructive aspect of what was happening particularly keenly. He opposed the restructuring at some cost to his own health. Eventually he collapsed and later took early retirement. The Trinity House Pilots were disbanded in 1988 and they became employees of the Port of London Authority. I was working in New Zealand at that time and I remember the telephone call which came through from my mother, telling me what had happened and that he was badly shaken, but recovering. I remember speaking to him on the phone then and listening to the change in his voice helplessly and angrily. The title of the poem preserves the date of the phone call.

October 1st, 1988

Oh father, father gone among –

There is no harbour
and no sunlight
the span of the bridge is not there

Absence is a fixed date
in memory. Desire
is what we'll have to plan for now

There are no big ships
and no-one there to pilot them
we have disembarked –

Then
Suddenly you are speaking to me
and your voice
is not as it was.

Years later, after my father's retirement, my parents came to New Zealand for my wedding. Later, they visited again, and one bright day I drove them and our young son up to Auckland for a day's outing. We took young James to Kelly Tarlton's Underwater World, where we were hurled around on a conveyor platform under a transparent tube beyond which sharks and stingrays, fish of all kinds, were swimming overhead. After this, we had lunch in the seafood restaurant upstairs, and then walked out onto the headland to take a look at the harbour and the port of Auckland. Suddenly my father's eyes grew keen and his attention quickened. What happened next is recorded in the poem, and there is a sequel I'll come back to in a moment.

Port of Auckland

New Zealand Herald, 9.12.98: SHIPPING NEWS
From Sydney, 12.45pm: *Botany Bay*

Too far away, the ship swings in: my father's eyes
clock funnel, colours, movement, recognise *Bay Line*
I listen to his pilot's voice quietly speak Scots:
I stood upon that bridge, along the London River
some other morning now too old to detail, but
close enough to know. The sun and stars and moon
make charts and keep us moving. On Orakei Jetty by
Bastion Rock, a spit reaching out from the near,
that seems as though it might go on forever into
blatant sky, we pause before we back along, and
watch her slowly clear. My young son takes my hand
suddenly scared, the wind he walks into will rise.

After my parents returned to Scotland from this visit to New Zealand, my father told me that he had been at a reunion lunch organised by the Honourable Company of Master Mariners, and in the course of a conversation, found that he was talking to the man who had been the captain of the *Botany Bay* on the very day we saw her come into the port of Auckland. It was a chance in a million, we reckoned, because that was her last voyage. She was overnight in Auckland then sailed north to Japan to be sold for scrap. We had noted her passing at exactly the moment prior to her being broken up, at the end of her life.

Alexander Riach in St Paul's Cathedral, Dunedin

When my parents visited New Zealand for the first time, we discovered another strand of the story of my father's family history. My father's grandfather's brother was Alexander Riach, who was born 12 March 1834, emigrated to New Zealand and died in Dunedin in 1921. He was known as the Red Stonemason of Dunedin and helped design and build the altar in St Paul's Cathedral there.

The Dunedin Riachs

The story behind the discovery was that after my parents were married in 1956, my father's father received a letter from a Mrs Jean Campbell in Oamaru, asking him whether my father might be related to a relative of her own. She had seen the marriage notice in the *Weekly Scotsman* newspaper, and noticed that my father's name was exactly the same as the name used by members of her own relations, in the Scottish tradition of naming the grandson after his grandfather. Thus my son James is named after my father, and my own middle name is Scott, after my father's father's middle name. My parents kept Jean Campbell's letter and when I went to New Zealand in 1986, they gave it to me and asked me to make contact if I was ever in Oamaru. In my first year I didn't go there, but when my parents visited, we drove around the whole country, north and south islands, and they found themselves in Oamaru. My mother took the letter to the local police station and asked if they could tell her whether Mrs Campbell was still alive. The policeman checked a computer and told her that yes, indeed, she was very much alive, having recently passed her driving test. He could not give out her address but in the circumstances, he felt it reasonable to give my mother her phone number. My mother dialled the number.

'Hello?' said the voice at the end of the line.

'Hello. Is that Mrs Jean Campbell?'

'Yes, it is.'

'I wonder, does the name James Alexander Riach mean anything to you?'

There was a pause, then: 'Indeed it does! I wrote a letter about James Alexander Riach in Scotland more than thirty years ago and I haven't had a reply yet!'

My mother told her that they were there in person, Mrs Campbell invited them over, and the stories of the Riachs of New Zealand and the Riachs of Scotland began to be told, and make connections.

2 O. R. D.
Oamaru
South Island
New Zealand.
Oct. 14th 1959

Mr Riach
Dear Sir.

You will wonder who this letter is from + what it is all about. Well, I have been reading some (back dated) copies of the "Weekly Scotsman" given to me by friends who have recently settled here in New Zealand. In one of these papers I saw the account of your son's wedding. In New Zealand the name Riach is very uncommon but was my name before I married — Your sons name, James Alexander, are family names with us being handed down from my great grandfather to my brother. I wondered when I saw your name if perhaps our great Grandfathers were related. Going back a long way, arn't I? My great Grandfather Alexander Riach came from Rothes in Morayshire, arrived in New Zealand in January 1861 in the ship "Lady Egidia" with his wife + son (my Grandfather)

Letter from (Mrs) Jean Campbell

179

He was a stone mason by trade
+ my grandfather carried on the good
work. That is a brief account &
if by chance there is any connection
I would be very pleased indeed to
hear from you

yrs sincerely

(mrs) Jean Campbell.

(Great Grandfathers were brothers)
            Jim says -

Living in New Zealand as I did from 1986 to 2001, the seafaring connections back to Captain Cook and the Maori voyagers in the South Pacific were a constant source of interest and curiosity. Cook sailed into the Pacific for a purpose, to observe the Transit of Venus. At one point he wrote: 'I had ambition not only to *go* farther than any man had been before, but as *far* as it was *possible* for a man to *go*.' How far does one go to find love? The metaphoric power of that phrase, the Transit of Venus, stayed with me a long time before I came to write this poem.

Transit of Venus

Geometry of contact
                            as 6 hrs past
made compasses from one point to
another,
            an angle therefore figured indicates
the only world
                            particular endurance and the will
allows to make love happen in,
                            by virtue of the calculated distances
the means it takes forever to devise
lifetimes of devotion, against the measureless tides

        : like the clock at Beauvais
a mechanism made to signal, chimes each hundred years:
a sound its maker could hear only once. And that's
like love, or children, all
                            that mortal being matters.
The Transit of Venus:
        The way love moves
                            the orbits of the sun and other stars.

In 2007, I had the good fortune to travel on a sailing yacht, the *Correyvreckan*, skippered by Glen Murray, who was kind enough to write the introduction to this book, on a one-week cruise around the Inner Hebrides. We sailed from Oban south to Gigha, up through the Sound of Islay and into harbour at Erraid, anchoring in David Balfour's Cove, familiar to readers of Robert Louis Stevenson's *Kidnapped*. We sailed to Iona, out to Staffa, then north and round to Tobermory, before returning to Oban.

These two poems were written about that trip.

The Element

A flight of dolphins shoaling by
as if it were not sea but sky.

A minke's tall black curving fin
slides anti-clockwise, right back in.

Two shearwaters skate the air
wings as sharp as razors there.

And up the sound of Islay in Force 7, the ketch and bodies angled
in sea that makes all light, water, vision, muscle, tense, taut, tangled.

At anchor, the curious seals rise now and then, and keek,
their rounded heads are bullet points that punctuate the week.

The element of the sea was there, all right, but it was the human connection across time, through memory, that was brought home to me when we anchored in Tobermory Bay.

At Tobermory

The ripped sail needs patching.
Needle and thread.

Held taut
as the strong hands direct, point and draw.
A criss-cross of hatchings contract and connect.

It will hold, and the sail keep its belly
when filled.

    Needle and thread:
returning to the harbourfront,
the narrow streets and closes traced behind it –
my father's childhood haunts –
the tree-filled hills rise up beyond the curve of bay, the row
of brightly-coloured shops, hotels and bars, the pier –
pausing once or twice to say hello
to older friends from long ago.

Then back to the sea once again.

– Held taut, the mind will point and draw.
The thread cross-hatched and holding.

One of the earliest and most shocking experiences for my father
was his first encounter with India. I remembered this when I went
to Kolkata for the first time, in 2012. I arrived early in the afternoon
and decided that it would be an experience worth having to hire
a taxi to drive me into the city, across the Howrah Bridge where
I could see the Hoogly River, and return me to my hotel. The
second of the three poems that came from that trip is entirely in
my father's words.

Passages from India: Three Poems

I: Calcutta

My father has been here before.
It is a dream. What happens to you later
cannot be predicted. Kiddapore Docks.
The sky is full of the bridge, this fretworked grid of steel,
and all its weight is clamped across the city.
Its 27,000 tons bear down, and yet
the life below is far from being crushed: it crushes,
crashes round itself, its market is unending.

Washing, getting filthy, drinking, eating,
passing it all, the fires in the dusk in the alleys
and streets, the orange light in the dun living twlight.
Thin men leaning on a wall or bench or car
bonnet, fat men in the restaurants, women
walking quietly, their eyes alert and looking.

It is as if the men are watching, the women,
different, looking. Unobtrusively. The children have not
settled, yet, just how to use their eyes.
They are full of surprise, and active.

My father has been here before me.
At the start of the world and after the end of the war,
the Merchant Navy Clan Line brought him in,
standing on the bridge, a young man, First Mate,
leaning into the world for the first time then.

Brown smoke from the traffic,
the permanent grit in the air, in your hair,
the sticky asphalt of the road,
the broad drifts of dust,
the refuse risen in contours,
hardened into the earth itself,
and the salvaging, the raking through
the residue for sustenance –
what power might be derived
from broken things.

Ubiquitous as dust,
intimate in nostrils and lungs,
felt in the eyes and tasted in the mouth,
it is about the place, a sense
of what we owe ourselves,
a debt unpaid, an uncrossed cheque,
a state of how becoming
may be coming clearer.
What cruelty, distressful, is the fact,
yet there is this presence here,
this marvellous disturbance.

Beyond the rage and bloodshed, as there is,
is also this persistence:
the mutiny of the tyrannised,
the fact of validation.

What world is this, to hold us all?
What happens to you later
cannot be predicted. But it is not a dream.
These lives are real, the river, bridge, the people,
the commerce and the currents take us all.

II: James Alexander Riach

'The biggest thing that hit me I remember. I was eighteen years old.
We sailed into Cochin on the Malabar coast, with an Indian crew,
to change the crew. It was 1948. We'd to pay them all off,
and get a new crew. It was a Clan Line ship, the *Clan MacAulay*.
We were carrying wool from Australia, bound for Liverpool and
      Glasgow.
I was a cadet. (My earliest voyage, the first voyage I did
was all apples: 192,000 cases of apples, from Tasmania,
bound for Liverpool, and Glasgow. That was maybe, 1947.)

'But it was India that hit me. It hit me like the biggest wave I ever saw,
higher than a house, that swept away the lifeboats, and I was on the
      bridge,
looking out. That was how India hit me: poverty, the way the people
did what they did, on shore, like nothing I had ever seen,
eighteen years old, from Stornoway on Lewis and from Mull and
      Dingwall.
Nothing I had ever seen was like that. That's how I remember it.'

III: Kolkata

Howrah Bridge –
      moving, seen through traffic
between buildings, above people –
people on the pavements and crossing the roads, in crowds

and crowds, people in the buildings,
passing through doorways, entering, leaving,
every building now this taxi drives me past, and above,
in the windows, moving, and above the buildings, the bridge,
moving, and beyond the traffic, buses, people, packed with people,
and inches from the taxi window, the sides of buses, scarred and
    dented,
moving in the constant noise of horns and shouting, talking, moving,
rising over the people and the traffic and the buildings,
the bridge rising into the dust and mist and brown and grey
polluted air come down to rest on its highest structures, them
rising into it, it meeting them, folding into them,
inhumanly high and intimate,
Howrah Bridge –
    So the taxi takes me through the streets
and the maze of routes, the labyrinth,
these canyons of Kolkata,
till we come to the great whirlpool,
the gulf of the circular roads around
the central railway station. And now the bridge is close,
looming. The traffic stops at the lights and all
the engines are switched off
as the crowds pour off the pavements, over the roads,
to the other sides, torrential surge of movement,
sensitized as people are, but caught among and moving along
these huge, irreversible tides.
        And now to cross,
the bridge takes us on and into its structure.
The current drives us all along its span.
For a minute or two or three, perhaps,
the vast tranquil shape of the river below
is seen in the twilight of pink, ochre, delicate
grey, and I roll the window down
as the January air rolls in, cool and
suddenly clean, and the ships along the river
can be seen in their separate places,
moving or at anchor in the commerce
of their cargoes and the purpose
of their destinations. —
And then it's over, we're over,

and the curve round the road takes us back,
to return, crossing once again,
back through the city, back to
where we came from.
    Howrah Bridge behind us
and the river far below.

There are always other stories, waiting to be told. When I returned to Scotland after that trip to India in January 2012, news of the obscene excesses of bonus payments made to a number of British bank managers was on the front pages of all the newspapers. The distance between the extremes of the poverty I encountered in Kolkata and the extravagance of money in the world of these immoral individuals was a reminder of how shocking such things should be to anyone who believes that all human lives are intrinsically valid. When I talked to my father about this, we agreed that the bankers of London seemed to possess less dignity than the people of Kolkata.

That sense of human dignity, the validity and value of life, are what I think my father's stories and my poems might help with, across all their encounters. I hope they do.

The Last of the Trinity House Pilots

The Last of the Trinity House Pilots

James Alexander Riach

Further Reading

Richard Baker, *The Terror of Tobermory: An Informal Biography of Vice-Admiral Sir Gilbert Stephenson, KBE, CB, CMG* (London: W.H. Allen, 1972)

John Clarkson, Roy Fenton and Archie Munro, *Clan Line: Illustrated Fleet History* (Preston: Ships in Focus Publications, 2007)

John Masefield, *The Conway* (London: Heinemann, 1933; revised edition 1953)

John Masefield, *The Sea Poems* (London: Heinemann, in association with The Marine Society, 1978)

Glen Murray, editor, *Scottish Sea Stories* (Edinburgh: Polygon, 1996)

Iain Crichton Smith, *Mr Trill in Hades, and Other Stories* (London: Victor Gollancz, 1986)

James Alexander Riach

Further Reading

Richard Baker, *The Terror of Tobermory: An Informal Biography of Vice-Admiral Sir Gilbert Stephenson, KBE, CB, CMG* (London: W.H. Allen, 1972)

John Clarkson, Roy Fenton and Archie Munro, *Clan Line: Illustrated Fleet History* (Preston: Ships in Focus Publications, 2007)

John Masefield, *The Conway* (London: Heinemann, 1933; revised edition 1953)

John Masefield, *The Sea Poems* (London: Heinemann, in association with The Marine Society, 1978)

Glen Murray, editor, *Scottish Sea Stories* (Edinburgh: Polygon, 1996)

Iain Crichton Smith, *Mr Trill in Hades, and Other Stories* (London: Victor Gollancz, 1986)

Lightning Source UK Ltd.
Milton Keynes UK
UKOW03f0144291013

219979UK00005B/51/P